Mangrove Microorganisms

Biodiversity and Biotechnology

The Authors

Dr. Hrudayanath Thatoi (b.1963) obtained his M.Sc; M.Phil and Ph.D. from Utkal University and presently working as Professor in Department of Biotechnology of North Orissa University. Dr. Thatoi has more than 12 years of teaching and 22 years of research experience. His areas of teaching and research includes microbiology, molecular biology and biotechnology. Dr. Thatoi has implemented several research projects funded by UGC-DAE Govt. of India, DST Govt. of Odisha and Forest Department, Govt. of Odisha. He has been associated in conservation and management of mangroves of Odisha coast through implementation of NORAD and ICEF projects through M.S. Swaminathan Research Foundation, Chennai and studied ecology and pollution problems of Bhitarkanika and Mahanadi delta by implementation of two small projects funded by Forest Department, Govt. of Odisha. He has served as co-editor for the "Mangrove conservation and Restoration" published by M.S. Swaminathan Research Foundation, Chennai in 2002. Dr. Thatoi has published more than 225 research papers in national and international journals, proceedings of conference and book chapters. Besides he has authored 7 books including two text books; one on microbiology and other on Practical Biotechnology. So far, he has guided 12 Ph.D. scholars and several M. Tech and M.Phil students. He has vast experience on the areas of mangrove conservation and restoration and exploration of mangrove microbial diversity from Odisha coast along with evaluation of their biotechnological potentials.

Dr. Rashmi Ranjan Mishra (b. 1983) obtained his M.Sc. and Ph.D. from North Orissa University and is presently working as Associate Professor in the Department of Biotechnology, MITS School of Biotechnology, an affiliated college of Utkal University, Odisha, India. He is teaching microbiology, animal biotechnology, biochemistry in UG and PG level. He obtained his Ph.D. on the topic entitled "Microbial diversity from mangroves of Bhitarkanika: A study on genotypic, phenotypic and proteomic characterization of some predominant bacteria". His research focused on assessment of microbial diversity from mangroves and evaluation of their biotechnological potentials. Dr. Mishra has more than 8 years of teaching and research experiences in the field of microbiology and biotechnology. Dr. Mishra has published more than 30 research papers in international and national journals as well as book chapters. He has received the best young scientist award in the year 2017 from the "Society of Biotechnology and Bioinformatics, Bhubaneswar.

Dr Bikash Chandra Behera (b. 1984) obtained his M.Sc. and Ph.D from North Orissa University and is presently working as a Technical Assistant in NISER, Bhubaneswar. His Ph.D topic is "Chracterisation and evaluation of biotechnological potentials of some soil bacteria from mangrove environment of Mahanadi delta Odisha. His research focused on the study of ecology and diversity of microbes from mangroved environment and evaluation of their biotechnological potentials. Dr Behera has published more than 20 research papers in national and international journals as well as several book chapters.

Mangrove Microorganisms

Biodiversity and Biotechnology

Prof. (Dr.) Hrudayanath Thatoi
Dr. Rashmi Ranjan Mishra
Dr Bikash Chandra Behera

2018

Daya Publishing House®

A Division of

Astral International Pvt. Ltd.

New Delhi – 110 002

© 2018 AUTHORS

ISBN 9789387057708 (International Edition)

Published by : **Daya Publishing House®**
A Division of
Astral International Pvt. Ltd.
– ISO 9001:2015 Certified Company –
4736/23, Ansari Road, Darya Ganj
New Delhi-110 002
Ph. 011-43549197, 23278134
E-mail: info@astralint.com
Website: www.astralint.com

PREFACE

Mangroves represent highly dynamic and fragile ecosystems yet they are the most productive and biologically diversified habitats of various life forms including plants, animals and microorganisms. The mangrove ecosystems support genetically diverse groups of aquatic and terrestrial organisms. Due to rich sources of nutrients, mangroves are rich in microorganisms. Microorganisms forms integral part of the mangrove ecosystem. They help in recycling and transformation of various nutrients and thus make the mangrove ecosystem more productive. All microbial forms such as bacteria, fungi, cyanobacteria, microalgae, macroalgae, fungus like protists and actinomycetes have been reported in this ecosystem. The microbial community of this ecosystem contributes to the bio-geochemical cycle and nutrient transformation by various mechanisms like degradation of complex molecules, ammonification, nitrification and denitrification as well as carbon flux and methane utilisation.

Mangrove microbial diversity exhibits novel biotechnological applications due to its remarkable adaptation to such a unique swampy, saline and partially anaerobic environment. The biotechnological potentials of mangrove microorganisms include characterization of novel enzymes with useful applications to human life such as in agriculture, industry, medicine and production of other value-added products like drugs, therapeutic proteins, antibiotics, vaccines and diagnostic tools etc. Actinomycetes isolated from mangrove habitats are a potentially rich source for the discovery of anti-infection and anti-tumor compounds, and the agents for treating neurodegenerative diseases and diabetes. Besides their enzyme and antibiotic production, mangrove microbes harbour tremendous potential for bioremediation and environmental protection. Phospahte solubilizing bacteria (PSB) inhabiting in the mangrove soil can directly utilized for biofertilizer production and can be used in agriculture for plant growth improvement. Mangrove leaf litter contains high amount of cellulose which favors the growth of cellulose degrading bacteria. The cellulase enzyme from these cellulose degrading bacteria can be used for many biotechnological applications.

The particular conditions of a mangrove and adaptation of bacterial species to such conditions, represents an important source of biotechnological potential resources to be exploited. A phylogenetic and functional description of microbial diversity in the mangrove ecosystem has not been well addressed to the same extent as that of the other environments. The microbial diversity and distribution in a mangrove would improve our understanding of bacterial functionality and their interactions found in that ecosystem. Notwithstanding the existing knowledge of microbes and microbial processes, we are still at the base of microbial diversity, which needs to be explored for the judicious and gainful utilization of this nature's treasure.

In the light of this, it was considered necessary to collect as much information as possible on study of microbial diversity in mangroves environment in general and Odisha in particular. The present publication entitled "Mangrove Microorganisms: Biodiversity and Biotechnology" has been prepared with a hope that it generate interest among readers and researchers to learn about the ecology, microbial diversity and its biotechnological potentials.

We are thankful to all those who have helped us directly or indirectly in the preparation of this book and Astral International Pvt. Ltd., New Delhi, India for publishing this book.

Prof. (Dr.) Hrudayanath Thatoi
Dr. Rashmi Ranjan Mishra
Dr. Bikash Chandra Behera

Contents

1

INTRODUCTION

Mangroves are a diverse group of salt tolerant plant communities found in tropical and sub-tropical intertidal regions of the world between latitudes 24^0 N and 38^0 S. These mangroves are commonly found in sheltered coastal areas subjected to tidal influences which receive rainfall between 1,000 to 3,000 mm and temperature ranging from 26-35 ^{0}C. The mangroves have been variously defined in literature. The term "mangrove" refers to an assemblage of tropical trees and shrubs that grows in the intertidal zones. Thus, mangrove is a non-taxonomic term used to describe a diverse group of plants that are all adapted to a wet, saline habitat. Some environmental conditions such as cold and high wave energy prevent them from growing, while mangroves are adapted to other conditions like salt, flooded soils, tidal inundation etc. Their adaptation to salt water allows them to flourish where no other trees can survive. Terms such as mangrove community, mangrove ecosystem, mangrove forest, mangrove swamp and mangal are used interchangeably to describe the entire mangrove community. The term 'mangal' was commonly used in French and in Portuguese to refer to both forest communities and to individual plants. Mangrove can grow quite well in fresh water, but their special ability to regulate salt allows them to out-compete most other trees in saline condition in the tropical and subtropical tidal environments. Mangrove prevent excessive amount of salt from damaging their tissues by restricting the amount of salt that can enter through roots, expelling salt crystals through transpiration, growing a thick cuticle to restrict salt absorption, or by salt excreting glands. Since mangrove plants live part of their life under submerged or flooded condition, they have adaptations such as prop roots and pneumatophores, for supplying roots in wet soils with oxygen. Mangrove's environment pose reproductive challenges due to flooding and anaerobic condition, hence reproductive adaptations have been developed in mangroves. The mangroves develop floating seeds (propagule) that germinate while still attached to the mother plant. Actually, the mangrove embryo doesn't develop into resting seed-stage, but continues its development into a plant while still on the tree, which is referred to as a propagule instead of a seed. This process is called vivipary germination and it gives the mangrove propagule a head start before they are dropped into the harsh coastal environment.

Mangroves represent highly dynamic and fragile ecosystems yet they are the most productive and biologically diversified habitats of various life forms including plants, animals and microorganisms. These are often called as "tidal forests", "coastal woodlands" or "oceanic rainforests". Mangroves grow normally along the land-sea interface, bays, estuaries, lagoons, backwaters and in the rivers, reaching upstream up to the point where the water remains saline (Qasim, 1998). They flourish in diversified habitats such as core forests, litter-forest floors, mudflats, water bodies (rivers, bays, intertidal creeks, channels and backwaters), and adjacent coral reefs as well as in sea grass ecosystems. A rich biodiversity is observed in the mangroves with plants and animals, which are irreplaceable and form a good genetic treasure house. The mangrove ecosystems support genetically diverse groups of aquatic and terrestrial organisms. Mangroves play important role in maintaining the ecology and livelihood of the in coastal region. People who live adjacent to mangrove ecosystem derive direct and indirect benefit from mangrove ecosystem.

1.1. Importance of Mangroves

Mangrove wetlands are a multiple use ecosystem that provides protective, productive and economic benefits to coastal communities (Table 1). Mangroves contribute to the stabilization of the shoreline and prevention of shore erosion. They serve as a barrier against storms so as to lessen damage to coastal land and residents. The dense network of supporting roots and breathing roots give mechanical support to the tree and trap the sediments. Without mangroves, all silt will be carried into the sea, where turbid water might cause corals to die. Mangrove trees act as sinks, which concentrate pollutants such as sewage, toxic minerals, pesticides, herbicides, etc. The mangrove ecosystems support rich diversity of organisms and serve as the nursery and breeding ground of several marine fauna like prawns, crabs, fishes and molluscs. They enhance the fishery production of nearby coastal waters by storing nutrients and detritus. Through shedding of their leaves mangroves tend to remove salt accumulated in them. However, most land plants, except the halophytes, are incapable of such tolerance to salty environments. The shed leaves of mangroves turn into detritus, which is colonized by fungi and bacteria and this detritus turns into a valuable pool of nutrients. The detritus is consumed by variety of bivalves, shrimps and fishes, many of which migrate into the mangrove areas for better feeding and protection. Birds are a prominent part of most mangrove forests and they are often present in large numbers. They use mangrove environments as breeding and feeding grounds. Mangroves are also rich in animal diversity which includes terrestrial animals such as tiger, wild bear, snakes, crocodiles, shrews, otters, civet cats, monkeys, eagles, dears, kites, king fishers, waders, monitors, spiders, insects etc. The Royal Bengal Tiger is one of the unique resident species of mangroves of Sunderbans. Similarly, salt water crocodile is dominant animal in mangroves of Bhitarkanika. Mangroves sustain the ecological security of the coastal areas as well as livelihood security of the thousands of fishing people of estuarine and coastal villages. Traditionally, local communities in mangrove ecosystems collect fuel wood, harvest fish and other natural resources. Mangroves are relatively well known for their floral diversity which is comprised of only 65-69 species of vascular plants which have several specific adaptations to the dynamic coastal environment

(Kathiresan and Bingham, 2001). Nevertheless, because mangroves grow where nothing else grows, they are always useful **(Fig.1)**, even where they cannot be managed as productive forests.

Table 1: Some important uses of mangroves

Ecological uses	Economical uses	Social uses
Erosion control and stabilization of shoreline, soil formation	Breeding ground of fishes and birds, wildlife habitat	Education
Protection from storm and tsunami/ hurricanes	Source of livelihood of fishing communities, Support shrimp industries	Ecotourism
Indicator of climate change	Charcoal production	Livelihood Support
Control coastal pollution, Nutrient Cycling	Timber, firewood and other forest produce	Local employment
Water quality management, biofiltration	Harbour large number of microbial community	Agriculture
Carbon sequestration, Climate Change Mitigation	Fodder production	Traditional medicine
Produce detritus	Refuse of marine organisms	Aesthetic value

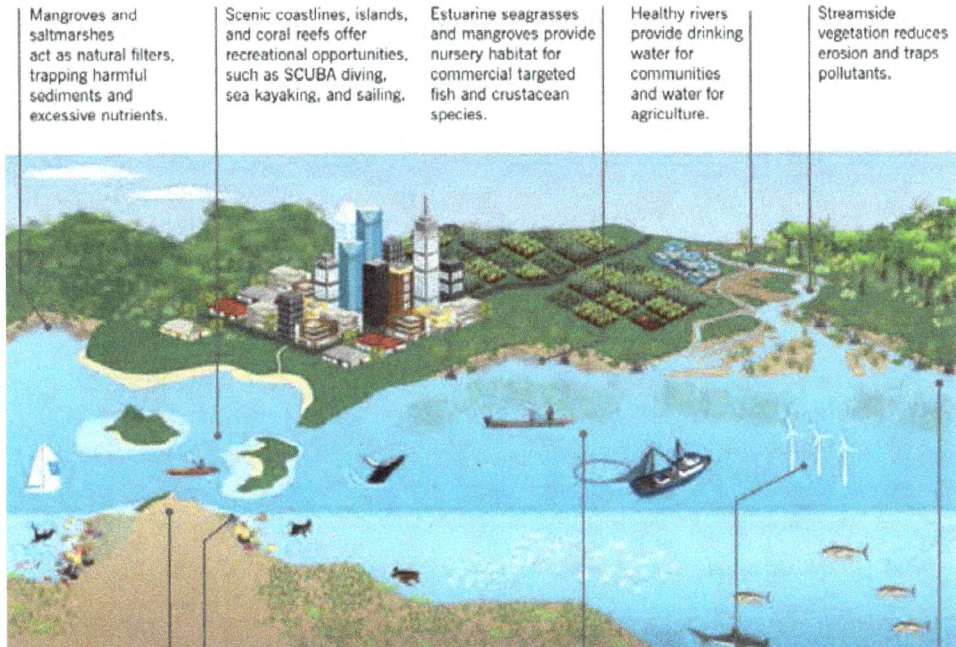

Mangroves and saltmarshes act as natural filters, trapping harmful sediments and excessive nutrients.

Scenic coastlines, islands, and coral reefs offer recreational opportunities, such as SCUBA diving, sea kayaking, and sailing.

Estuarine seagrasses and mangroves provide nursery habitat for commercial targeted fish and crustacean species.

Healthy rivers provide drinking water for communities and water for agriculture.

Streamside vegetation reduces erosion and traps pollutants.

Fig. 1 Some important uses of mangrove

1.2. World Mangroves

The total mangrove area was estimated to be 137,760 km² distributed in 118 countries and territories in the tropical and subtropical regions of the world.

Approximately 75% of world's mangroves are found in just 15 countries (Table-2). The total mangrove area accounts for 0.7% of total tropical forests of the world. The largest extent of mangroves is found in Asia (42%) followed by Africa (20%), North and Central America (15%), Oceania (12%) and South America (11%) (Giri *et al.*, 2011). Asia has large mangrove areas because the environmental conditions are especially favorable for mangrove growth. This can be attributed to the regions highly conducive environment for the growth of mangrove forests, characterized by such qualities as a humid climate, high rainfall and a number of rivers with large deltas supplying fresh water and sediments. The important ones include the Ayeyarwady delta in Myanmar (Burma), the Mekong in Vietnam, and the extensive deltaic coastline along southern Papua in Indonesia. Moreover, this region is also known as the global center of mangrove diversity, with 51 species, which is 71% of the total mangrove species found all over the world.

Recently, world mangrove report by FAO updated the list of countries with mangrove forests as 124 countries possessing the mangrove forests as against. The differences in the list of countries are found as countries with small areas of mangroves, were excluded from many of the earlier studies. The FAO report also estimates the total mangrove area at 15.6 million hectares. Mangroves may grow as trees or shrubs according to the climate, salinity of the water, topography and edaphic features of the area in which they exist.

Table 2: The 15 most mangrove-rich countries of the world (Giri *et al.*, 2011)

Sl. No.	Country	Area (ha)	% of global total	Region
1	Indonesia	3,112,989	22.6	Asia
2	Australia	977,975	7.1	Oceania
3	Brazil	962,683	7.0	South America
4	Mexico	741,917	5.4	North and Central America
5	Nigeria	653,669	4.7	Africa
6	Malaysia	505,386	3.7	Asia
7	Burma	494,584	3.6	Asia
8	PapuNew Guinea	480,121	3.5	Oceania
9	Bangladesh	436,570	3.2	Asia
10	Cuba	421,538	3.1	North and Central America
11	India	368,276	2.7	Asia
12	Guinea Bissau	338,652	2.5	Africa
13	Mozambique	318,851	2.3	Africa
14	Madagascar	278,078	2.0	Africa
15	Philippines	263,137	1.9	Asia

Mangroves may found as isolated patches of dwarf stunted trees in very high salinity and/or disturbed conditions or as lush forests with a canopy reaching 30-40 meters in height under suitable environmental conditions. Mangrove forests consist of relatively few plant species in most diverse habitats, 30-40 species can be present; however, in many places only one or a few occur. Mangrove species are not diverse in terms of plant species. The exact number species is still under discussion and ranges from 50-70 according to different classifications (Saenger *et al.* and Davie, 1983; Lugo and Snedaker, 1975; Aksornkoae *et al*; 1992) with highest species diversity found in Asia, followed by eastern Africa. In general, mangroves have been categorized into two groups: true mangroves and mangrove associates (Selvam *et al.*, 2004). True mangroves grow only in mangrove environment and do not extent into terrestrial plant communities. In the other hand, plants which can grow in other coastal environments and also within the mangroves are considered to be mangrove associates. This group includes terrestrial plants as well as pure halophytes. As per global assessment, there are 73 species and hybrids from 20 different families are considered as true mangroves. Out of 73 true mangrove species 38 are also considered as "core" mangrove plants. Families of *Rhizophoraceae* contribute majority of the core species. While mangrove trees are the predominant vegetation in most areas, mangrove plants also include ferns, several shrubs and a palm (Spalding *et al.*, 2010).

1.3. Indian Mangroves

India has three types of mangrove habitats, namely deltaic, backwater-estuarine and insular. The deltaic mangroves are luxuriantly present on the east coast (**Fig.2**) (Bay of Bengal) where the gigantic rivers make mighty deltas such as the Gangetic, the Mahanadi, the Godavari and the Cauvery deltas. The backwater-estuarine types of mangroves exist along the west coast (Arabian Sea) and are characterised by typical funnel-shaped estuaries of major rivers (Indus, Narmada, Trapti etc) or occur in the backwaters, creeks and neritic inlets. The insular mangroves are present in Andaman and Nicobar Islands, where many tidal estuaries, small rivers, neritic islets and lagoons support a rich mangrove flora. Mangrove wetlands in India are also classified based on their spatial and temporal variations of environmental factors (Selvam, 2003). These are (i) tide-dominated e.g. the Sunderbans and Bhitarkanika mangroves and (ii) river dominated e.g. the Godavari, Krishna, Pichavaram and Muthupet mangroves.

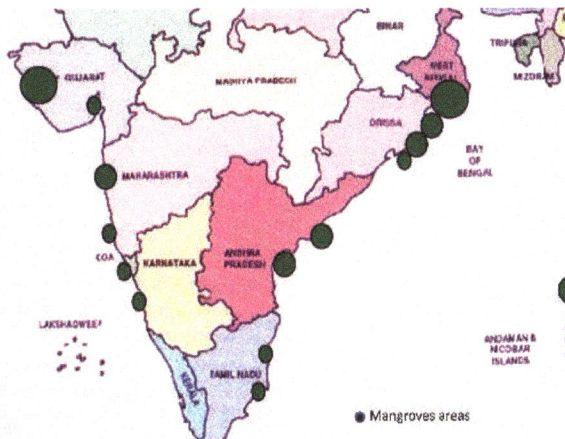

Fig. 2 Distribution of mangroves in India

The current assessment (FSI, 2015) shows that mangrove cover in India is 4,740 sq. km which is 0.14 percent of the country's total geographical area. The very dense mangrove comprises 1,472 sq km (31.05 percent) of the total mangrove cover; moderately dense mangrove is 1,591 sq km (29.35 percent) while open mangroves constitute an area of 1,877 sq km (39.60 percent). There has been a net increase of 112 sq km in the mangrove cover of the country as compared to 2013 assessment (FSI, 2015). This can be attributed to increased plantations particularly in Gujarat state and regeneration of natural mangrove areas. The mangrove vegetation is about 57% in the East coast and 23 % in the West coast. The Bay Islands (Andaman and Nicobar) account for 20% of the country's total mangrove area. India harbours some of the best mangrove forests of the world along the river mouths along the east and west coasts (Table-3) as well as, the Andaman and Nicobar Islands in the east coast. Geomorphic settings of the mangrove wetlands of the east coast of India are different from that of the west coast. East coast has a smooth and gradual slope which provides large areas for colonization of mangroves whereas the west coast has a steep and vertical slope (Thatoi and Biswal, 2008). The coastal zone of the west coast is narrow and steep in slope due to the presence of the Western Ghats (Gopal and Krishnamurthy, 1993). Mangroves are much more extensive on the east coast of India due to nutrient rich alluvial soil formed by several rivers which also supply fresh water along the deltaic coast. Along the east coast, the Gangetic Sunderbans in West Bengal is the largest mangrove cover in India. Besides, large rivers like Mahanadi, Brahmani-Baitarani, Krishna, Cauveri and Godavari also harbour major mangroves in their estuarine regions along east coast. The west coast mangroves are dominated by estuarine back waters which include the coastal area of Gujarat, Maharashtra, Goa, Karnataka and Kerala.

Table 3: Mangrove swamps in India (FSI, 2015)

Sl. No	States/UTs	Very dense mangrove	Moderately dense mangrove	Open mangrove	Total mangrove
1	Andhara Pradesh	0	129	238	367
2	Goa	0	20	6	26
3	Gujarat	0	174	933	1107
4	Karnataka	0	3	0	3
5	Kerala	0	5	4	9
6	Maharashtra	0	79	143	222
7	Odisha	82	95	54	231
8	Tamil Nadu	1	18	28	47
9	West Bengal	990	700	416	2106
10	A&N Islands	399	168	50	617
11	Daman and Diu	0	0	3	3
12	Puducherry	0	0	2	2
Total		**1472**	**1391**	**1877**	**4740**

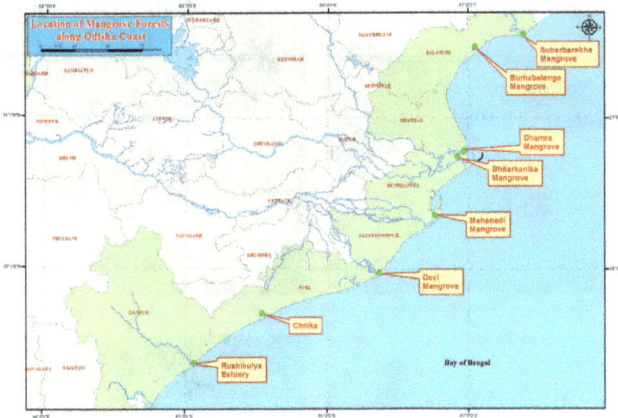

Fig.3 Distribution of Mangrove in Odisha

1.4. Mangroves of Odisha

Odisha is a maritime state located in the east coast of Indian peninsula and having a long coast line of 480 km. The mangroves of Odisha are distributed in six major river deltas such as the Budhabalanga, Subernarekha, Brahamani-Baitarani (Bhitarkanika), Mahanadi and Devi river mouths (**Fig.3**). Odisha has comparatively larger mangrove habitats in the country due to the nutrient rich alluvial soils of the river deltas formed by these rivers (**Fig. 3**). The coast also has a smooth and gradual slope which provides larger areas for colonization of mangroves. As per an estimate, the state has a 231 km^2 total mangrove areas distributed over five coastal districts that include 82 sq. km very dense mangrove, 95 sq. km moderately dense mangrove and 54 sq. km open mangrove cover (State of Forest Report, 2015). The mangrove forests of Odisha occur in discontinuous patches in the districts of Balasore, Bhadrak, Kendrapara, Jagatsinghpur, Puri and in the fringes of the Chilika lagoon. At present the mangroves of Odisha are mainly confined in a continuous belt for a distance of about 160 km from the mouth of Dhamara river in the north to Devi river mouth in the south through the deltaic formations of the major rivers like Brahmani, Baitarani, Mahanadi and their tributaries.

Table 4: Mangroves of Odisha (FSI, 2015)

SL No	District	Very dense mangrove	Moderately dense mangrove	Open mangrove	Total Area in Km^2
1	Balasore	0	0	2	2
2	Bhadrak	0	9	21	30
3	Jagatsinghpur	0	2	6	8
4	Kendrapara	82	84	24	190
5	Puri	0	0	1	1
Total		82	95	54	231

Fig.4 Mangrove forest of Odisha coast

The Mangroves of Bhitarkanika (Brahmani-Baitarani river deltas) are very rich and abundant due to the enforcement of a wildlife protection act, since the estuary was declared as a sanctuary in 1975. Mangrove vegetation in Mahanadi delta region between Barunei mouth to Mahanadi mouth (Paradip) is fragmented and is in degraded state mostly due to the construction of Paradeep Port (1965), development of industries and diversion forests for human settlement, paddy fields and aquaculture ponds. Further south of Paradeep, sparse mangrove vegetation occurs along the coast from Mahanadi mouth to Devi mouth.

Mangroves occurring to the north of Dhamara mouth of Bhitarkanika up to Chudamani in Bhadrakh district coast and also on Budhabalanga and Subernarekha mouths in the Balasore district are highly degraded state. Mangroves occurring in the Rusikulya river mouth and fringes of Chilika lagoon have been disappeared. However, some mangroves have been planted in the inter tidal areas of Sipakuda and Arakhkuda of Chilika lake by th local communities with support of NGS and Forest Department, Government of Odisha. The details of the mangrove occurring in river deltas are discussed below.

1.4.1. *Mangroves of Subarnarekha* and *Budhabalanga*

Mangroves of river mouths of Subernarekha and Budhabalanga along the Bay of Bengal in Balasore district of Odisha, at present are in highly degraded state owing to heavy biotic pressure. These mangroves forests are coming under Baripada Forest Division. Mangrove forest at Subarnarekha river mouth has been notified as Protected Reserve Forest (PRF) vide notification 63002 dt. 15/9/1980 over an area of

563 ha as Bichitrapur Bani PRF. It is under Jaleswar range of Baripada division and Bhograi C.D. Block of Balasore district. Subernarekha forms an estuary complex at its mouth composed of mangrove trees, bushes, salt marshes, mudflats and sandy beaches at the extreme north eastern part which opens into Bay of Bengal. Long mud flat (19 km) besides the narrow creeks, surrounded by mangrove bushes and salt marshes is found between Talsari to Kirania. Creeks are covered by patchy mangroves (Mitra and Pattanayak, 2003). The Subarnarekha estuary is endowed with a few mangrove species of *C. decandra* and *B. parviflora*. Concurrently, *A. ilicifolius*, *E. agallocha*, *D. trifoliata*, etc. are found as associate species. The ground flora and sand dune vegetation are completely absent in this region except some herbaceous species like *C. arenarius*, *C. platystylis* and *F. ferruginea* etc.

River Budhabalanga originates from the Similipal hills of the Mayurbhanj district and enters Balasore near Kalyanpur of Remuna Block. Balaramgadi is the place where the river Budhabalanga meets the sea, which is 2 kms away from Chandipur. Balaramgadi is a large intertidal zone facing the Bay of Bengal with predominance of sandy and clay sediments. The Budhabalanga estuary harbours only the devastated as well as degraded mangrove forests. There exist naturally growing mangrove patches approximately over 15 ha of area near the mouth of river Budhabalanga. Vegetation is represented by shrubby elements and stunted forms of tree species. The notable mangrove species are *Avicennia officinalis, Bruguiera gymnorrhiza, Sonneratia apetala, Excoecaria agallocha, Rhizophora mucronata, Bruguiera cylindrica, Ceriops decandra, Acanthus ilicifolius, Caesalpinia nuga, Myriostachya wightiana, Suaeda maritima, Porteresia coarctata*, etc. This region is completely devoid of ground flora and sand dune vegetation. About 50 ha of degraded lands have been planted with seedlings of *Avecennia* sp., *Excoecaria agallocha, Ceriops decandra, Rhizophora mucronata, Bruguira gymnorrhiza, Aegiceras corniculatum*, etc. by the Odisha Forest Department during 2001-2002 .

1.4.2. Mangroves of Bhitarkanika

The mangroves of Bhitarkanika, between the Dhamara mouth to Barunei mouth on the east coast, has been notified as Bhitarkanika sanctuary (672 Sq.km.) in 1975 in order to give protection to the endangered salt water crocodile (*Crocodilus porosus*) along with the flora and fauna of this region. It is largest mangrove forest of Odisha. The core area of the sanctuary (145 sq.km) is notified as National Park in September, 1998. This stretch of mangrove is the only area in Odisha, which is relatively well preserved. Bordering the Bhitarkanika sanctuary/National Park, an area of 1435 Sq. km. (out of which 1408 sq.km. is sea to a width of 20 kms. form the coast) has been declared as Gahirmatha marine wildlife sanctuary in September, 1997. It also covers two reserve forest blocks of Mahanadi delta mangroves comprising 27 Sq.km.

Bhitarkanika sanctuary area has been designated as a 'RAMSAR SITE' (i.e., Wetland of International importance) during the 8th meeting of the contracting parties held at Valencia from 18-26 November 2002. In Odisha this is the 2nd wetland of international importance under Ramsar Convention (after Chilika) and is one of the 19 such sites in the country. Mangrove vegetation of Bhitarkanika is very rich and diverse. It harbours as many as 65 mangrove species and their associates which is maximum in Indian context. The dominant species in this region are *Sonneratia apetala, Avicennia officinalis, A. alba, Heritiera fomes* and *H. littoralis* etc. Occurrence of three species each of *Avicennia, Heritiera, Sonneratia, Rhizophora* and *Xylocarpus* and four species of *Bruguira* is significant for this forest. Besides, faunal diversity in Bhitarkanika is also very high. The mangroves of Bhitarkanika support 174 species of birds including migratory ducks and geese and largest concentration of bareheaded geese in India. There are 26 species of mammals, 5 species of amphibians, 44 species of reptiles and several species of fishes, and numerous species of invertebrates in these mangroves. The important carnivores include the endangered fishing cat (*Felis viverrina*) and leopard cat (*Felis bengalensis*). Further, these mangroves constitute the last habitat for sizeable population of endangered reptiles such as the saltwater crocodile (*Crocodylus porosus*), Indian python (*Python molurus*), King cobra (*Ophiophagus Hannah*), water monitor (*Varanus salvaor*) and the Olive ridley sea turtle (*Lepidochelys olivacea*). Bhitarkanika is also rich in prawn fisheries. The mangrove and associated forests provide the subsistence requirement of timber, fuel wood, tannin, honey and thatch roof for the local people and fodder for the local communities (Chadah and Kar, 1999).

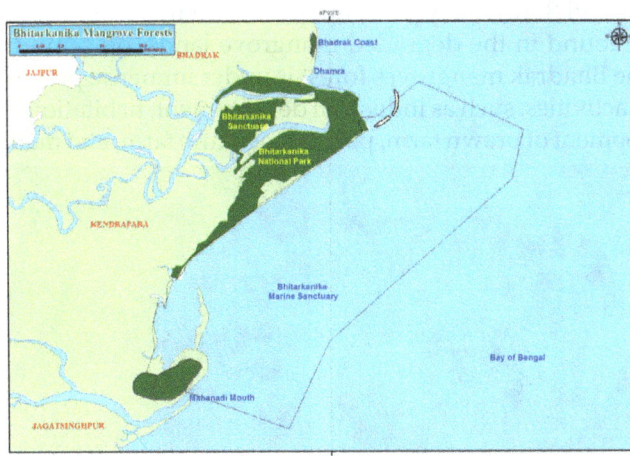

Fig.5 Map of Mangroves of Bhitarkanika Odisha

1.4.2.1 Mangroves of Bhadrak district in the North of Bhitarkanika

Bhadrak District has 50 km long coastline from south bank of river Kansabansa to north bank of river Dhamara intercepted by natural creeks in many places. *Avicennia marina* is the dominant mangrove species followed by *Avicennia officinalis, Avicennia alba, Acanthus illicifolius, Excoecaria agallocha* etc in this mangrove forest. However, mangrove of Bhadrak coast is now under immense pressure due to various anthropogenic activities. Mangrove forest of Bhadrak coast, north of Bhitarkanika ranges from river Kansabansa south bank up to north bank of river Dhamara all along the 50 k.m long coastlines. Bhadrak coast had great importance which has Dhamara, Chandabali, Chudamani mouth areas. The mangrove forests (Bani forest) protect Bhadrak coast from damage due to high tide. The Bhadrak coastline with mangrove forest acts as bio-shield against oceanic disaster and protects the villages and agriculture from time immemorial. Bhadrak coast is also home to Horse shoe crab which is an endangered species. Many valuable fish, crab, prawn species are available in Bhadrak coast that sustain the livelihood of coastal community and strengthen to state economy. Bhadrak coast is endowed with unique type of mangroves forest i.e. Fringing mangrove forest influenced by daily tidal range. Mangroves forest exists continuously along the coast intercepted by natural creeks. The whole stretch of mangrove forest is locally known as BANIPAHI. Due to its long exposure to purely marine conditions, it supports *Avicennia marina* (O-BANI) species. The forests are dominated by *Avicennia marina* 99.9% and few numbers of other species like *Avicennia officinalis, Avicennia alba, Acanthus illicifolius, Exocaecaria agallocha* are also present. In the Dhamra estuary, degraded mangrove forests are seen along with some of the associates in small patches. The true mangroves are *A. officinalis, C. decandra, S. apetala*, which exhibited very stunted growth giving the appearance of small bushy trees. The associates consist of *D. trifoliata, P. paludosa, T. populnea*, etc. *P. coarctata* is usually abundant here, where as *M. wightiana* is found comparatively in lesser abundance. Both species have a great role in checking soil erosion. The ground flora is very poor. But pure formations of *S. portulacastrum* and *S. maritima* are found in the degraded mangrove lands, on semi-dried mudflats. Now-a-days the Bhadrak mangroves forest is under immense pressure due to high anthropogenic activities, such as industrial development, habitation by Bangaldeshi refugee, development of prawn farm, port, aquaculture farm, and many more **(Fig.6)**.

Fig.6: Bhitarkanika mangrove forest

1.4.3. Mangroves of Mahanadi delta

Mangroves of Mahanadi delta extends from south-eastern boundary of Mahanadi river mouth to Hansua (a tributary of Brahmani) in the north; from northeastern end of Mahanadi river up to Jambu river in the west. Mangrove ecosystem in the Mahanadi river delta occupies about 38.56 sq. km extending between Hansua river mouths in the north to Paradeep port in the south. Mangroves of Mahanadi delta is an arcuate type of delta formed by Mahanadi river and its tributaries. Mahanadi mangrove ecosystem is the second largest mangrove ecosystem of Odisha next to Bhitarkanika which is in degraded state although some parts of this mangrove forest show very rich mangrove vegetation as seen in Bhitarkanika. Mangroves of Mahanadi delta comprise of 34 true mangroves, 9 obligate mangroves and 42 mangrove associates. The dominant species in Mahanadi delta are *Avicennia officinalis*, *A. marina*, *Sonneratia apetala*, *Excoecaria agallocha* and *Rhizophora mucronata* etc. Due to operation of biotic factors of high magnitude and habitat loss in the Mahanadi delta many taxa have moved towards threatened categories.

1.4.4. Mangroves of Devi river mouth

Mangroves are found in Devi river mouth further sourth of Mahanadi in the district of Jagatsingpur. Devi river arises from river Kathjodi, a tributary of river Mahanadi drains into Bay of Bengal forming a tidal estuary with meandering creeks and channels. The Devi river reaches Bay of Bengal 70 KM south of the mouth of Mahanadi river. The mouth of the Devi river also serves as a nesting ground for Olive Ridley turtle. Lakhs of turtles come to this region every winter for breeding. The sea facing side of the coast has long sand bars covered with casuarina plantation taken up by forest department. On the inward side small creeks, streams and their branches have mudflats which are suitable for mangrove vegetation. The estuary of Devi river is almost devoid of typical mangrove elements mainly due to habitat destruction connected with human settlement and paddy cultivation in the areas. Moreover, the ecological conditions have been changed due to formation of sand bars which considerably checked inundation. Only patches of *Acanthus ilicifolius*, *Tamarix troupii*, *Excoecaria agallocha*, *Myriostachya wightiana*, *Phoenix paludosa*, etc. are found in denuded condition. At places old stumps of *Avicennia* sp. and *Heritiera fomes* are reminiscent of the existence of past mangrove vegetation. In mud flats, *Suaeda maritima*, *Suaeda monoica*, *Sesuvium portulacastrum* and *Fimbristylis ferruginea* are common elements. In recent past, efforts have been made by State Forest Department and M.S. Swaminathan Research Foundation, to introduce a number of species i.e., *Avicennia officinalis*, *Sonneratia apetala*, *Aegiceras corniculatum*, *Ceriops decandra*, *Bruguira gymnorrhiza*, *Rhizophora apiculata*, *Avicennia marina*, etc. in the form of plantation. In total, 12 true mangroves and 4 mangrove associates are present in Devi river mouth.

1.4.5. Remnants of mangroves of Rusikulya river mouth

Apart from the above six river deltas, mangroves were also present in the Rusikulya river mouth which are however not found at present. The Rusikulya river opens to the Bay of Bengal near Ganjam town. In the Rusikulya estuary, no significant mangrove formation is observed at present. Only pure stands of *S. maritime* an associate of mangrove are very common. Among the sand dunes, *S. littoreus*, *S. trilobatum*, and *I. pes-caprae* are very common. The mass–nesting beach (rookery) for Olive Ridley turtles in the southern Odisha coast is located at Rushikulya river mouth.

1.4.6. Mangroves at Chilika

As evidenced from the palynological studies of sediments, in the past, the margin of the Chilika lake and the undisturbed islands like Badokuda, Sanokuda, and Ghantasila etc. were endowed with the unique eco-system of mangroves. Haines (1921-1925) and Narayan Swami and Carter (1922) have enumerated some mangrove species and their associates in fringes of Chilika. But in the course of time, the mangroves have been replaced by tidal scrub jungles and only some mangrove associates are present due to the change of the eco-climate conditions coupled with the ruthless cutting of trees by the local inhabitants to cater for their various needs. The vegetation is totally devoid of the true mangrove species and only a few associates are encountered in the denuded regions. Such species are *S. persica, C. ceylanica* and *P. pinnata* etc. The other species of this category are *A. corniculatum, A. tetracantha, Colubrina asiatica, C. inerme, P. aculeata, C. retusa, Meytinus emarginatus, Carissa spinarum, Opuntia stricta, P. suberosa* etc. *C. quadrangularis, P. capensis, C. roxburghii, C. zeylanica, Tylophora indica, S. acidum, D. trifoliata and M. oblongifolia* are quite abundant. The non mangrove trees of this region are *C. adansonii, L. tetraphylla* and *Streblus asper*. In the FSI report of 2013, the Chilika mangroves forest is not included. Mangroves forest is re-introduced in an area of 15 hectare close to the new mouth in the outer Channel of Chilika ascent to Sipakuda village since 2004 through community participation facilitated by PREPARE (NGO), FD and CDA Government of Odisha. Coastal communities of Chilika have planted mangroves on 40 hectares land in intertidal areas of Arakhuda (old mouth of Chilika lake) with technical support of Integrated Coastal Zone Management Project (IndCZM). The common species planted are *R. Apiculata, A. marina, A. officinalis, A. albaba, B. gymnorrhiza, S. apetala, K. candel*, (Sipakuda) and *Rhizophora, Excoecaria agalocha. A. marina, K. candel* (Arakuda).

1.5. Threats to Mangrove

Despite their values, mangroves are amongst the most threatened ecosystems world-wide, subject to over-exploitation, pollution, and conversion (Farnsworth and Ellison, 1997). Globally, high population pressure in coastal areas has caused the conversion and decline of mangroves. These, mangroves are generally undervalued, overexploited, and poorly managed (Ewel *et al.*, 1998a). Yet, their importance to humans, wildlife, and global carbon balance is paramount (Walters *et al.*, 2008; Kristensen *et al.*, 2008). Some experts have even cautioned that the long-term ecosystem services that mangroves provide will likely be lost within the next century as mangrove areas become smaller and more fragmented (Duke *et al.*, 2007). Mangrove forests rank among the most threatened of coastal habitats, particularly for developing countries (Field *et al.*, 1998; Saenger *et al.*, 1983). Once occupying around 75% of tropical coasts and inlets, now the mangroves are restricted to few pockets. Today, less than 50% remain and of this remaining forest, over 50% is degraded and not in good form. A figure of 1% decline per year has been given as the conservative estimate for the Asia-Pacific region (Ong, 1995). However, accurate estimates of global deforestation rates of mangroves are yet unavailable. Anthropogenic pressures on mangroves ecosystem are high. Among others, mangroves are clear felled and reclaimed for agriculture and aquaculture, urban expansion, developmental activities.In the developing countries pressures for fuel wood, poles, fodder etc. often exceed sustainable levels. Because of high calorific value people are destroying mangroves for firewood, charcoal and timber collection. Mangrove wood is highly suitable for chipboard and paper industry for which mangrove trees are cut down. In particular the large-scale conversion of mangrove forests to ponds for shrimp aquaculture is an underestimated problem (Naylor *et al.*, 2000; Dahdouh-Guebas *et al.*, 2002b; Primavera, 2005). Not only direct or destructive anthropogenic effects such as clear felling, but also indirect impacts such as changes in hydrography have proved detrimental to mangroves (Dahdouh-Guebas *et al.*, 2002a).

Another threat to mangroves is diversion or alteration of freshwater flow into them. Mangroves are well established in areas where there is good amount of fresh water inflow and where regular tidal flow occurs. Embankment construction and siltation at the river mouth obstruct tidal water flow into mangrove swamps. Reduction in fresh water and tidal water inflow increases the salinity of mangrove areas, also results in poor growth and regeneration of mangrove plants. For example, the mangroves of Pichavaram, South India are largely dying due to hyper salinity and other associated factors such as increasing temperature, poor precipitation and poor flushing of mangrove soil by tidal waters (Selvam *et al.*, 2002). Similarly, in Sundarbans, the reduction in fresh water inputs, resulted in reduction in population of species such as *Heritiera fomes* and *Nypa fruticans* (Bhatt and Kathiresan, 2011).

In most mangrove areas invasive species disrupt the ecological balance and dynamics of the mangrove ecosystem. For example, in Tamil Nadu and Andhra Pradesh, the rapid invasion of *Prosopis* species (Baruah, 2005) can be considered for invasive species. In Sundarbans, colonization of the twiner *Derris trifoliate* and other aquatic weeds *Eichhornia crassipes* and *Salvinia* in mangrove water negatively affecting the natural flora of mangrove ecosystems (Bhatt and Kathiresan, 2011).

In addition, climate change poses a threat to mangrove ecosystems (Gilman *et al.*, 2008). Climate change is one of the most important environmental issues impacting mangroves in India. It results in increase in temperatures, rising sea level, increasing the frequency of tropical storms and tsunamis. Due to sea level rise mangroves tend to move landward, but human encroachment prevents this and consequently, the width of the mangroves decreases.

The mangroves of Odisha have been severely affected by the felling of trees for fuels and timbers, the spread of urbanization, industrialization, aquaculture of prawns, and settlement of refugees. The mangrove vegetation particularly on the coast of Odisha (the Mahanadi delta, Devi estuary, Rusikulya estuary, Subarnarekha estuary, Chilika lagoon, etc.) has been highly denuded due to severe biotic pressure such as human settlement, construction of port and factories, prawn culturing and now a days only the remnants of the past vegetation in the form of shrubby elements are seen. Due to therapeutic as well as other uses, plants have been over-exploited ruthlessly. Unlike the other states of India, the coastal belt of Odisha has experienced severe vegetational devastations leading to rapid shrinkage of many rare and endangered plant communities like mangroves. Due to such devastations, the coastal belt has faced severe oceanic cyclones with high wind velocity which costs many lives.

1.6 Conservation and Management of Mangroves of Odisha

Mangroves play critical role in the economy of the people. It provides multiple benefits to the coastal communities apart from protection of the coastal environment. It is home to many wildlife and variety of aquatic organisms. Because of extraordinary ecosystem and environmental services of mangroves, their conservation has become most important. The conservation and management has been much similar to the management of any other inland forests. Govt. of India formulated the National Mangrove Committee (NMC) in the year 1976, as an advisory body to the Govt. of India to promote conservation of mangroves in India. Presently most of the Indian mangrove habitats enjoy the leglatative protection under the Indian Forest Conservation Act 1980 and the Wildlife (Protection) Act, 1972. The Indian mangrove receives special attention under the Wildlife Protection Act. 1972 as designated Marine and Coastal Protected Areas (MCPAs). Presently there are 31 designated MCPA in India, most of which involves mangrove habitats. Under coastal zone regulation, Indian mangroves are covered under the CRZ 1 and receive strong legal protection under environmental protection act, 1986. Further Costal Aquaculture Authority Act, 2005, also provides to the protection of mangroves. Several other legislative measures (Act & policies) such as Indian Fisheries Act, Indian Port Act, Coast Guard Act, Marine Fishing Regulation Act, etc. are also in place which directly or indirectly help in conservation of mangroves. India is particularly strong on policy front with adequate legal support for conservation of mangroves. Most of the mangroves are managed by state forest departments. However, presently many Indian mangroves are managed through community based co-management with forest department, scientific bodies, NGOs and other stake holders.

Mangroves of Orissa did not receive adequate protection in the past. Bhitarkanika forest area was under the control of the erstwhile Zamindari Forests of Kanika Raj till 1951. With abolition of the Zamindari system these lands vested in the State Government in 1952 (under Anchal Administration of Revenue Department). In 1957, the demarcated and notified protected forest blocks out of vested Zamindari forests were transferred to the control of the Forest Department. Mangroves of Bhitarkanika, Mahanadi and Devi river mouth are being managed by the Mangrove Forest Division, Rajnagar, mangroves of Subarnarekha, and Budhabalanga are managed by Baripada Forest Division and mangroves planted in Chilika are being managed by Chilika Forest Division, Govt of Odisha. With effect from 1.10.2003, the mangrove forests of Kendrapada and Jagatsinghpur Districts have been constituted into Rajnagar Wildlife Division, while the mangrove areas of Bhadrakh and Puri Districts are now part of Bhadrakh and Puri Wildlife Divisions.

Due to reclamation of mangrove forests for settlement of the immigrants, paddy cultivation, prawn culture, over-exploitation of woody mangroves, establishment of port and factories the mangrove forests have been denuded in most part of the state. However, in recent years much effort is being made by state Forest Department, Government of Orissa to not only conserve mangrove forests but also to restore the degraded mangrove areas. Apart from the Forest Department, some NGOs such as M.S. Swaminathan Research Foundation, Chenni, Prepare, Chennai are also involved in conservation and restoration of mangroves in the state. In view of much devastation of super cyclone hit Odisha coast during 29 and 30 November 1999, the ecological role of mangroves has been realized much better than the before and more emphasis has been laid on protection of this fragile forest ecosystem.

ECOLOGY OF MANGROVES

Mangrove is a complex ecosystem, composed of various inter-related elements in the land-sea interface zone. Both marine and terrestrial factors influence the mangrove ecosystem. They are the sources of primary productivity and involved in complex detritus based food webs (**Fig.7**). Mangroves have two components, mangrove forests and associated water bodies. A group of woody trees and shrubs that can grow well in saline water and water logged condition constitute the forest component. The associated water bodies comprised of tidal channels and canals that intersect mangrove forests. The mangrove forest and associated water bodies are together called mangrove wetland. Most of the mangrove wetlands are inundated by low saline water and sometimes by fresh water during the monsoons seasons and brackish water or sea water during other periods (**Fig.5**).

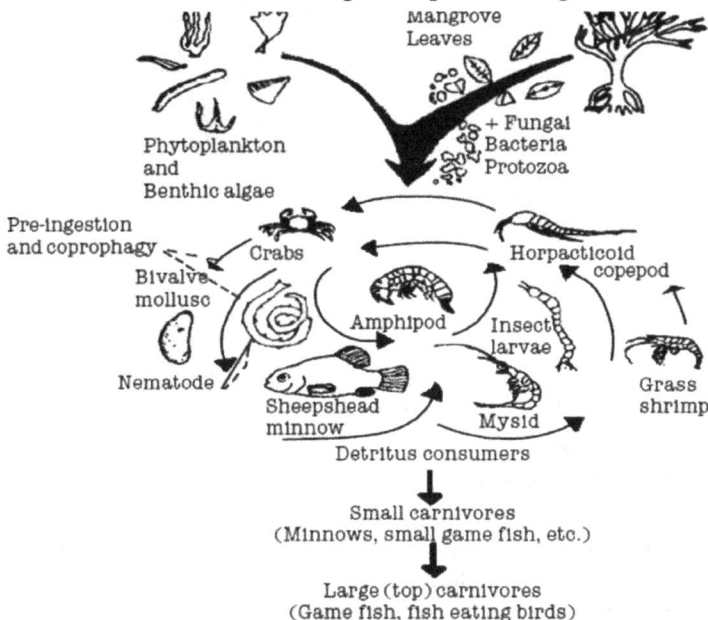

Fig. 7: Detritus based food web of mangrove (Selvam & Karunagaran, 2004)

Mangroves are naturally stressed ecosystems. Natural stressors include high soil salinities, tidal flows, storm tides, waves, excessive siltation or erosion and periodic hurricane or storm winds. All these factors contribute to strong environmental gradients that act as selective forces which in turn determine the distribution and density of species along the gradients. Depending upon species tolerance and the breadth of environmental gradients, vegetation zones of varying widths are evolved. The external factors believed to be the most important to mangrove function were tidal flushing, soil salinity, availability of fresh water and nutrients and climate. All these factors exhibit temporal and spatial variations and some are modified by the mangrove vegetation. Temporal variations include the tidal cycles, rain- fall cycles, seasonality of runoff and hurricane cycles. Spatial differences result from the location of the mangrove forest. Some grow closer to the sea than others and thus the relative influence of marine and terrestrial factors change. Climatic influences are important because they regulate freshwater availability, soil salinity, nutrient input from runoff and temperature. Mangrove growth is more vigorous in the continuously humid locations. In humid climates soil salinities are lower and nutrient inputs from the land are higher. In arid environments the excessive evaporation concentrates salts in the soil and terrestrial inputs are reduced. The amplitude of the annual temperature regime and the frequency of frost and conditions of low air temperatures become important which limits mangrove distribution in some places **(Fig.8).**

Fig. 8 A typical mangrove forest inundated with tidal flooding

2.1 Mangrove Ecotypes

Mangroves get tightly bound to the coastal environment in which they occur. Not only are they influenced by chemical and physical conditions of their environment but they usually help to create those conditions by themselves. They are found in a variety of tropical coastal settings like the deltas, estuarine areas, lagoons and fringes of the coral reefs. Topographic and hydrological characteristics within each of these settings define a number of different mangrove ecotypes. Six of the most common ecotypes include over wash, fringe, scrub, riverine, basin, and dwarf forests (hammock) (Lugo and Snedaker 1974; Twilley 1998) **(Fig.9)** and each types of forest are discussed below:

Overwashed mangrove: Overwashed mangrove forests are small mangrove islands frequently formed by tidal washings. In a mangrove wetland, the overwashed mangrove forest occurs in and adjacent to the bays on small islands, newly formed shoals and finger-like projections of larger land masses in shallow bays and estuaries. In most instances overwashed mangrove forests are situated almost perpendicular to tidal flow patterns and are overwashed with each high tide. The velocities of tidal currents are large enough to carry all loose debris into the inner bays. As these materials are not re-deposited by these retreating tides, there is an observable paucity of detritus in these forests.

Fringing mangrove: Fringing mangrove forest occurred along the borders (fringes) of protected shore lines and islands, influenced by daily tidal range. This type of mangroves is situated as thin fringes along the slightly sloping shorelines of waterways. Exposure on the shore line makes them vulnerable to erosion, strong wind and turbulent waves. Tides are the primary physical factor in fringe forests. However, tidal waters do not overwashed these forests. Tides export buoyant materials such as leaves, twigs and propagules from mangrove areas to adjacent shallow water areas. This export of organic material provides nutrition to a wide variety of organisms and provides nutrients for the continued growth of the fringing forests.

Scrub mangrove: Scrub mangrove forest in comparison to other four mangrove settings, forms dwarf mangrove settings and flat coastal fringes (Lugo & Snedaker, 1974). These are usually found in extreme environments where nutrients are limited or little fresh water is available. Trees are stunted and frequently not more than 1.5 m in height.

Reverine mangrove: Reverine mangrove forests are luxuriant patches mangroves existing along river and creeks, which get flooded daily by the tides. Salinity drops during the wet season, when rains cause extensive fresh water runoff. However, during the dry season, estuarine waters are able to intrude more deeply into river systems and as a result salinity increases. Nutrient availability in these systems becomes highest during periods when salinity is lowest which makes these systems highly productive. Thus it promotes optimal conditions for mangrove growth. This alternating cycle of high runoff/low salinity followed by low runoff/ high salinity suggests that riverine mangrove forests are the most highly productive of the mangrove communities.

Basin mangrove: Basin mangrove forests are stunted mangroves located along the interior side of the swamps, drainage depressions, channeling terrestrial runoff towards the coast. Basin mangrove forests occur on the landward side of a mangrove wetland. They are also called interior mangroves. These mangrove forests are protected against waves and often inundated only in frequently by the tides. Salinity in the mangrove forest varies greatly, depending on the circumstances. In areas of high rainfall or of substantial groundwater flow, salinity may be quite low. On the other hand, evaporation and removal of water by the mangrove trees may combine to raise salinity and in some areas soil may be distinctly hypersaline.

Hammock mangrove: Hammock mangrove forests are very similar to basin mangrove forest but differentiated from the basin mangrove forest as they are more elevated. Hence they are often isolated but still receive tidal influences. Due to high salinity and low nutrient mangroves in the hammock forest are always stunted.

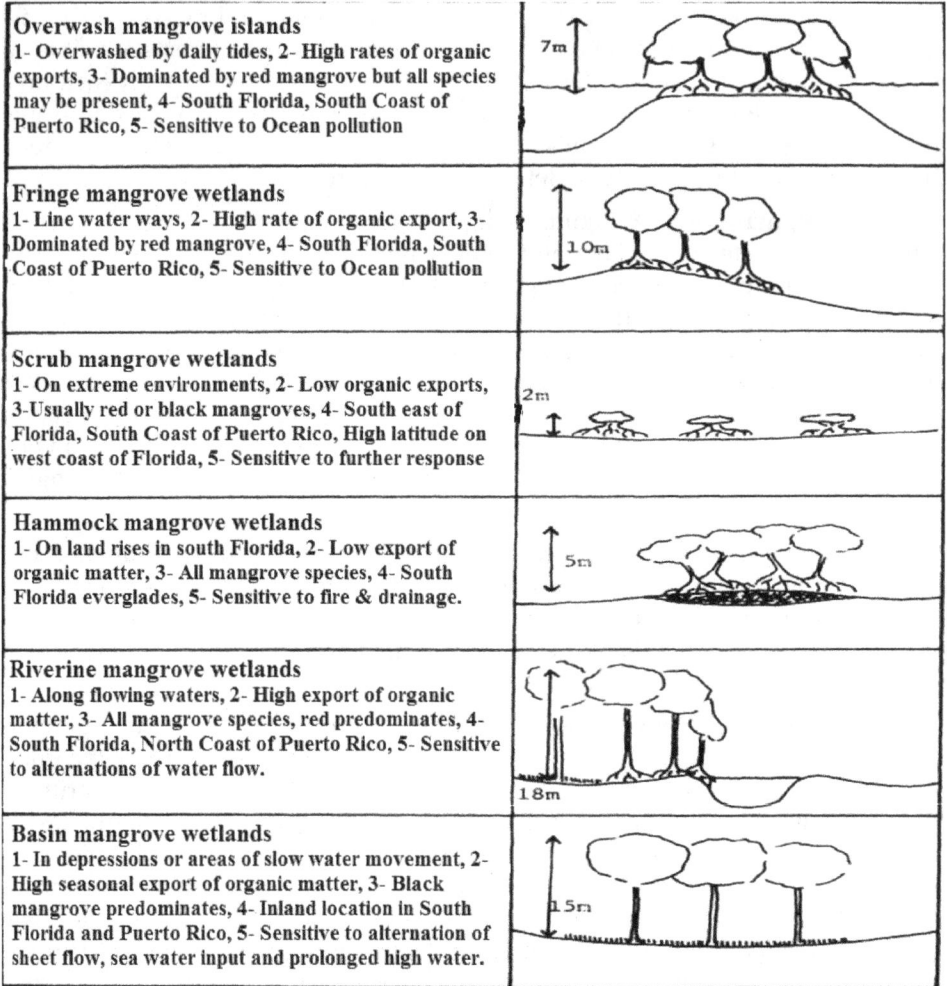

Overwash mangrove islands 1- Overwashed by daily tides, 2- High rates of organic exports, 3- Dominated by red mangrove but all species may be present, 4- South Florida, South Coast of Puerto Rico, 5- Sensitive to Ocean pollution	7m
Fringe mangrove wetlands 1- Line water ways, 2- High rate of organic export, 3- Dominated by red mangrove, 4- South Florida, South Coast of Puerto Rico, 5- Sensitive to Ocean pollution	10m
Scrub mangrove wetlands 1- On extreme environments, 2- Low organic exports, 3-Usually red or black mangroves, 4- South east of Florida, South Coast of Puerto Rico, High latitude on west coast of Florida, 5- Sensitive to further response	2m
Hammock mangrove wetlands 1- On land rises in south Florida, 2- Low export of organic matter, 3- All mangrove species, 4- South Florida everglades, 5- Sensitive to fire & drainage.	5m
Riverine mangrove wetlands 1- Along flowing waters, 2- High export of organic matter, 3- All mangrove species, red predominates, 4- South Florida, North Coast of Puerto Rico, 5- Sensitive to alternations of water flow.	18m
Basin mangrove wetlands 1- In depressions or areas of slow water movement, 2- High seasonal export of organic matter, 3- Black mangrove predominates, 4- Inland location in South Florida and Puerto Rico, 5- Sensitive to alternation of sheet flow, sea water input and prolonged high water.	15m

Fig.9 Structure of mangrove forest (Lugo and Snedaker, 1974)

2.2 Ecological Importance of Mangrove

In the coastal ecosystem like estuaries, mangrove ecosystem plays major role for diverse groups of organisms. Earlier ecologist regarded mangrove forests as unimportant, transitional communities with a low productivity. But most ecologists today view them as highly productive, ecologically important systems. Major roles of mangrove are:

1. Mangrove forests serve as protection for coastal communities against storms such as hurricanes. Mangrove communities protect shorelines during storm events by absorbing wave energy and reducing the velocity of water passing through the root barrier. In addition, mangroves protect intertidal sediment along coastlines from eroding away in harsh weather year round. The mangroves provide a natural wall, which is necessary in high impact natural disasters like tsunami.

2. Mangrove systems serve as habitat for many rare and endangered species of plants and animals. It is rich in food hence support many marine organisms such as fish, crabs, crocodiles, oysters and other invertebrates and wildlife such as birds and reptiles.

3. Because of the abundance of carbon and other nutrient contents from mangrove leaves and woods, the mangrove ecosystem harbours a large number of microbial communities, which have immense biotechnological, pharmaceuticals, agricultural as well as industrial application.

4. They protect and stabilize coastal zones by soil formation, nourish and nurture the coastal water with nutrients.

5. Mangroves act as filters for upland runoff.

6. Mangroves produce large amounts of detritus that may contribute to productivity in offshore waters.

7. Mangrove forests serve as nurseries and refuge for many marine organisms that are of commercial or sport value. Areas where widespread destruction of mangrove has occurred usually experience a decline in fisheries.

8. Mangrove forests are also important in terms of aesthetics and tourism. Many people visit these areas for sports fishing, boating, bird watching, snorkelling and other recreational pursuits.

2.3 Ecological Characteristics of Mangrove

Number of biotic and abiotic factors affects the mangrove plant growth (Kathiresan and Bingham, 2001). High salt, low temperature, drought and high temperature are common abiotic stress conditions that adversely affect plant growth and production (Mohammad *et al.*, 2009). Temperature fluctuation, salt stress and siltation are the major abiotic factor but the problem has been aggravated by human activity.

2.4 Limiting Water Loss

Because of the limited fresh water available in salty intertidal soils, mangroves limit the amount of water they lose through their leaves. They can restrict the opening of their stomata (pores on the leaf surfaces, which exchange carbon dioxide gas and water vapor during photosynthesis). They also vary the orientation of their leaves to avoid the harsh midday sun and so reduce evaporation from the leaves.

2.5 Nutrient Uptake

Mangrove soil is perpetually waterlogged hence; little free oxygen is available in the upper sediment. Anaerobic bacteria liberate nitrogen gas, soluble form nutrient (iron), inorganic phosphates, sulfides and methane, which make the soil much less nutritious. Pneumatophores (aerial roots) allow mangroves to absorb gases directly from the atmosphere and other nutrients such as iron from the inhospitable soil. Mangroves store gases directly inside the roots, processing them even when the roots are submerged during high tide.

2.5.1 Tidal gradient

Colonization of extensive mangrove forests depends upon coasts with a greater tidal range. A greater tidal range increases the intertidal area which encourages the growth of mangroves. Species that grow in the lowest intertidal zone successfully trap sediments. Over the time, the sediment builds up and new mangroves are able to invade and outcompete the colonizers.

2.5.2 Temperature

Mangrove plants do not adequately develop when annual average temperatures are below 19°C, which corresponds with the sea water isotherm of 20°C during the coldest period of the year (Alongi, 2008). While mangrove plants are intolerant to freezing temperatures both air and water temperatures may never decrease below 0°C. Optimal temperatures for mangroves are not only limited by cold temperatures but also by high temperatures because they hinder the tree settling. Since mangroves are found in the tropical and sub tropical region of the world, their optimum mangrove growth occurs only under tropical conditions where atmospheric temperature in winter season also greater than 20 C and the seasonal fluctuation does not exceed 5 C. Mangroves have been reported to grow in latitudes where the average sea surface temperature is 24°C. The optimum leaf temperature for photosynthesis is 28-32°C (Clough *et al.*, 1982; Andrews *et al.*, 1984). Any further rise in temperature may lead to spreading of only some species provided that the direction of ocean currents facilitates the dispersal of their seeds. The leaves of mangroves are sensitive to high temperature. Their photosynthetic capacity gets reduced, falling to zero at 38-40°C.

2.5.3 Salinity

Salinity is one of the major environmental problems affecting plants of different regions of the world. low osmotic potential of soil solution causing physiological drought, nutritional imbalances and specific ion toxicity or combination of all these factors are the major deleterious effects of salinity on plant growth. Mangroves plants are facultative halophytes. Consequently they can still grow and function well even up to a salinity of 90 ppt, but shown best growth when salinity fluctuates between 5 and 75 ppt (Krauss *et al.*, 2007). Salinity is one of the most investigated gradients in mangrove ecology which plays a vital role in their distribution of species, their productivity and growth (Twilley & Chen, 1998). They are fully capable of growing in fresh water. Mangroves species can tolerate higher salt concentration in comparison to non-mangrove plants (Ball and Pidsley, 1995). But in general, mangrove vegetation is more luxuriant in lower salinities (Kathiresan *et al.*, 1996). Mangroves are poor competitors under non-saline areas where freshwater marsh plants easily out-class them. Salinity at high levels also affects mangroves. Simple salinity fluctuations also have significant negative effects on the photosynthesis and growth of plants (Lin & Sternberg, 1993).

2.5.4 Rainfall and supply of freshwater

Freshwater inflows from the upper catchment influence estuaries and coastal waters in many ways. They are the major determinant of the environmental conditions in estuaries due to their impact on salinity gradients, estuarine circulation patterns, water quality, flushing, productivity and the distribution and abundance of many species of plants and animals. The availability of freshwater is an important factor for the development and growth of mangrove forests. Freshwater supply has often been indicated by the ratio of rainfall to evapotranspiration. Under the humid conditions, where the ratio exceeds 1, the mangroves grow luxuriantly. In arid climates, on the other hand, where the ratio falls below 1, the mangroves get stunted. High rainfall in humid areas leaches out residual salts from the mangrove soil and thus encourages the growth of mangroves (Jimenez, 1990). In the arid regions, barren salt flats often develop along the landward rim of the mangroves, due to poor leaching of salts from the soil due to low rainfall and this result in poor growth of mangroves because of high salinity of the soil.

2.5.5 Sediment characteristics

The sediments are characterised by the abundance of sand and silt with small amount of clay. The organic matter concentration ranges from 1.5 to 13.4% and are controlled by the particle size of the sediments. Enhanced concentrations of heavy metals in the surficial sediments were due to the abundance of greater surface area of fine particles, high organic matter content and flocculation process. These mangrove sediments may act as a sink for the metals derived from marine and fluvial processes. Different forms of available phosphorous in the mangrove sediments show spatial variation within the mangrove environment. One of the important functions of mangroves to environment is to provide a mechanism for trapping sediment, and thus the mangrove forests are believed to be an important sink of suspended sediment. The mangrove trees functioning as land builders as they catch sediment by their complex aerial root structure. The suspended sediment is introduced into coastal areas by river discharge, dumping of dredged material and re-suspension of bottom sediment by waves and ships. The mechanism of sediment transport in mangrove waters mostly based on the mechanism of hydrodynamic process rather than biological process (Ayukai and Wolanski 1997). The hydrodynamic process includes the asymmetry of the tidal currents, the baroclinic circulation and shear-induced destruction of flocs. The mangroves trap the suspended sediments during their transport based on tidal flows.

Three conditions are favour for mangrove formations. These are; (1) Open seashores where substratum is muddy and also protected, (2) Protected banks of estuaries and (3) Creeks which cut across the land mass. Among these three, protected banks of estuaries are the best suited locations for the mangroves.

One of the important factors in the mangrove habitat appears to be nutrient concentrations. Nutrient fluxes in these environments are closely dependent upon plant assimilation and microbial mineralisation (Alongi, 1996). Two major elements (N and P) are of great significance for the growth of mangroves. The concentration of dissolved inorganic nitrogen is generally low in tropical mangrove waters. Some microorganisms incorporate N_2 from the atmosphere and convert the nitrogen present in the soil into ammonium, which makes nitrogen available for the use of plants. Ammonia is the main form of inorganic nitrogen in mangrove soil. It combines with oxygen to form nitrite or nitrate by the process of nitrification, which is used by the plants. This process normally occurs in the root zone that releases oxygen. Like nitrogen, the concentration of inorganic phosphorus is also generally low in mangrove waters. It is normally available as a soluble salt that can readily be assimilated by the plants. Phosphate is efficiently adsorbed by the fine sediments of muddy areas rather than the coarse grained sediments. This is probably the reason for mangroves growing luxuriantly in muddy environments. Nutrient availability may limit growth and production in many mangals. The occurrence of sulphides is a characteristic feature of mangrove sediments and it influences the distribution of mangroves. In general, high sulphide levels can damage, reduce growth and cause high mortality of mangroves (Youssef & Saenger, 1998). Sulphide is formed due to reduction of sulphate. Clearing of mangrove forests or simple formation of canopy gaps can change the physical and chemical characteristics of the underlying soil, leading to anaerobiosis and an increase in the sulphide activity in the sediment. Heavy organic input can also increase sulphide production and the H_2S accumulation usually can kill the mangroves if their pneumatophores get covered by silt and are unable to transport oxygen to the rhizospheres.

2.5.6 Soil nutrient content in mangrove

Many environmental factors, including climate, geomorphology, hydrodynamics and soil characteristics, control the structure and function of mangrove ecosystems. Among all abiotic factors, edaphic properties, in particular soil nutrient status, have the most direct control on mangrove ecosystems (Boto *et al.*, 1984). Mangrove soils are typically saline, anoxic, acidic and frequently waterlogged. The delivery of nutrients in sediments and water during tidal inundation and sporadically in flood waters associated with cyclones and hurricanes provides significant sources of nutrients for mangroves (Lugo and Snedaker, 1974; Davis *et al.*, 2003). Generally, the mangrove sediments are reducing in nature (Alongi *et al.*, 2005; Prasad, 2005) and contain high amounts of organic matter (Morell and Corredor, 1993).

The availability of nutrients to mangrove plant production is controlled by dissolved and particulate nutrient pools in mangrove soils. These pools are regulated by tidal inundation, redox status and microbial activities, litter production and decomposition by mangrove plant (Boto *et al.*, 1984; Steinke and Ward, 1987; Boto, 1982; Holmer *et al.*, 1994). Therefore, nutrient status of soils varies significantly between mangrove ecosystems of different geographical locations. The redox state of the soil surrounding the mangrove roots is important for determining the nutrients available for plant uptake. In conjunction with the frequency and intensity of inundation, the redox state of soils is also influenced by the biota and the occurrence and abundance of mangrove roots. Thus, the redox state of the soil can be highly heterogeneous, facilitating a plethora of biogeochemical processes, which influence nutrient availability.

Large seasonal variations in soil nutrients could also be due to deposition of nutrient rich silt during the rainy seasons (Boto and Wellington, 1983). The mangrove sediments are anoxic and are also a source of microbially mediated greenhouse gases, e.g. CO_2, CH_4, N_2 and N_2O (Purvaja *et al.*, 2004).

2.5.7 Heavy metal content in mangrove soil

Anthropogenic inputs such as industrial activities, discarded automobiles, batteries and waste water discharge etc are the major sources of heavy metals that are found in mangrove soils. (Marchand *et al.*, 2006; Pekey, 2006; Shriadah, 1999; Bloom and Ayling, 1977). Biological degradation of heavy metals are much difficult, hence they are transferred and concentrated into plant tissues from soils which pose long-term damaging effects on plants. Heavy metal's cycling is a serious problem in mangrove environment due to toxicity and bioaccumulation as shown by many studies on mangrove environment (Shriadah, 1999; Tam and Wong, 1995). Due to their persistence, potential toxicity and bioavailability, heavy metals represent a major threat for mangrove environment. Heavy metals from incoming tidal water and fresh water sources are rapidly removed from the water body and deposited onto the sediments. Because of the capacity of mangrove to efficiently trap suspended material from the water column (Furukawa *et al.*, 1997) and the high affinity of organic matter for heavy metals (Nissenbaum and Swaine, 1976), mangrove sediments have large capacity to accumulate these pollutants (Tam and Wong 2000). In mangrove sediment many complex geochemical processes, complicated by great variability of major pore water parameters (pH, redox and salinity) take place. These processes vary considerably due to seasonal and spatial variations; more over reciprocal effect exist between plant species and sediment geochemistry (Merchand *et al.*, 2004). As a result, mangroves may act as a sink or source of heavy metal along tropical and subtropical coastal areas (Harbison, 1986; Tam and Wong, 2000) and are thus a natural laboratory to study metals distribution between dissolved and the solid phases (Merchand *et al.*,2012).

Although there have been a number of studies on heavy metal distribution in mangrove sediments in many countries (Lacerda *et al.*, 1993; Silva *et al.*, 1990; Mackey *et al.*, 1992), few data are available for mangroves in the subtropical regions, in particular, in India (Harbison, 1986; Lacerda *et al.*, 1993; Tam and Wong, 1995).

2.6 Nutrient Recycling in Mangrove Environment

Large quantities of organic matters from mangrove ecosystem are discharged to adjacent coastal waters in the form of detritus. These detritus serves as a nutrient source and is the base of an extensive food web. Transformation of nutrients from dead mangrove vegetation into sources of nitrogen, phosphorus and other nutrients are mainly carried out by the diverse microbial communities living in mangrove ecosystems which can be used by mangrove plants **(Fig. 10)**. In turn, plant root exudates serve as a food source for the microorganisms living in the ecosystem. The detritus can be defined as organic matter in the active process of decomposition.

It is rich in energy and contains a large active microbial population both attached and living free (Odum and Heald, 1975a). Microscopic examination of decomposing mangrove leaves reveals a complex community composed of bacteria, fungi, protozoa and microalgae (Odum and Heald, 1975b).

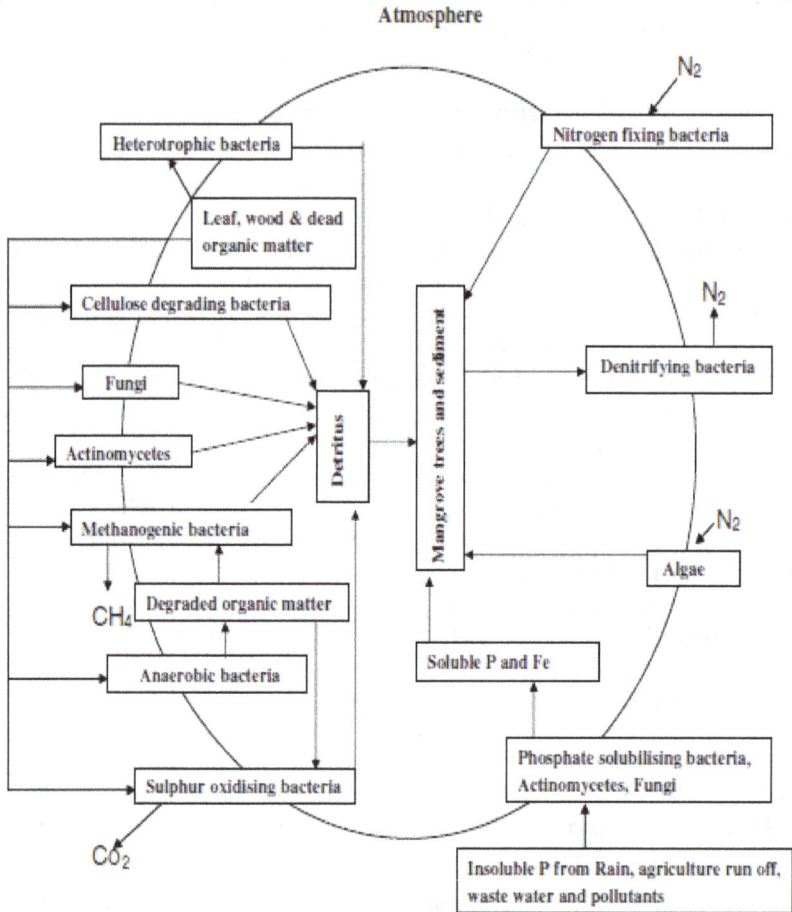

Fig. 10: Ecological role of microorganisms in mangrove environment

The growth and productivity of mangroves are strongly related to the benthic nutrient pools and nutrient transformations by the microbial decomposition of organic matter (**Fig.11**). Aerobic and anaerobic decomposition of these organic matters produce many oxidized (CO_2) and reduced products (H_2, NH_4, H_2S CH_4) which are further oxidized to give the end product of respective biogeochemical cycles. When organic matter is initially degraded aerobically, produces CO_2 which is the end product of carbon cycle. The H_2 produced from the degradation of organic matter, acts as electron donor in many biogeochemical cycles. It reduces the CO_2 to produce CH_4 by methanogenesis. CH_4 produced is again oxidized to produce CO or

CO_2, the initial ingredient for photosynthesis to produce the organic matter. Similarly the H_2S produced after the decomposition of organic matter are used in the sulphur cycle by the sulphur oxidizing bacteria to give the end product such as sulphate. The sulphate is again reduced by the process of assimilative sulphate reduction into organic compound (proteins) by the plants. Similarly ammonium (NH_4) produced after the decomposition of organic carbon compound is converted to nitrite and then nitrate by *Nitrosomonas* and *Nitrosococcus* through the process of nitrification, which is again available for direct plant uptake or anaerobically converted to N_2 by denitrification. The N_2 produced after denitrification is fixed by the nitrogen fixing bacteria and used up by the plant again to produce organic nitrogen compound. Unlike C, N, cycle the phosphorus doesn't involve the atmoshphere. Though organic 'P' compound include nucleotide, nucleic acid and phospholipids, ATP plays most essential role in living system but after the decomposition of the organic matter the phosphorus anion released are precipitated by the cation (e.g. Ca^{2+}, Mg^{2+}, K^+, Na^+) forming insoluble forms of aluminum, calcium or iron phosphates, all unavailable to plants. Thus microbes present in the ecosystem plays significant roles in solubilizing these insoluble phosphates and helps to avail these important nutrient for the growth and plant uptake.

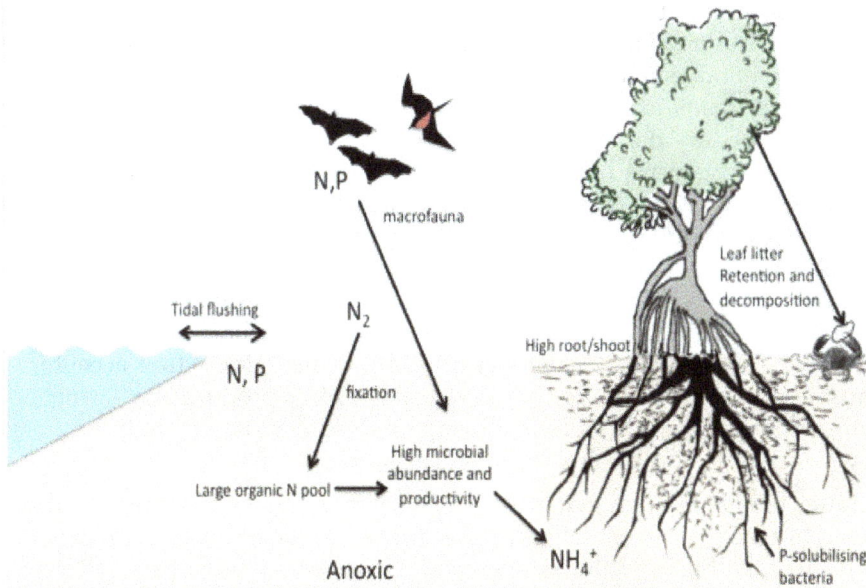

Fig. 11: Transformation of nutrient in mangrove soil (Reef *et al.*, 2010)

2.7 Carbon Content and its Cycling in Mangrove Soil

Mangrove ecosystems are rich in organic matter due to litter fall and the low rates of decomposition imposed by anoxic soils. Litter from trees (leaves, propagules and twigs) and subsurface root growth provide significant inputs of organic carbon to mangrove sediments (Alongi *et al.*, 1998). Mangrove forests fix, release and sequester more carbon by area than all other coastal ecosystem types (Alongi and Mukhopadhyay, 2014). Mangrove ecosystems, therefore, supply a substantially larger amount of carbon to the coastal waters and influence the global biogeochemical cycling of nutrients (Dittmar *et al.*, 2006). It has been reported that 11 % of the total organic carbon across the land–ocean interface in the tropics is of mangrove origin hence the carbon fixed by mangrove is potentially significant in the carbon biogeochemistry of the coastal zone (Jennerjahn and Ittekkot, 2002).

The degradation of organic matter in mangrove sediments is mediated by various microbial fermentation and respiration processes. A fraction of mangrove detritus escapes degradation and is permanently buried within the mangrove sediments or adjacent ecosystems. While some mangrove forests largely retain detritus within their sediments (i.e. as degradation or burial), others lose a major fraction of their net primary production to adjacent coastal waters mainly through tidal forcing as dissolved organic carbon (DOC), dissolved inorganic carbon (DIC), particulate organic carbon (POC) and particulate inorganic carbon (PIC) (**Fig. 9**). Because of the regular tidal flooding and draining in many mangrove forests, the material exchange with adjacent waters can be very efficient. Bacteria are responsible for most of the carbon flux in tropical mangrove sediments. The element carbon which forms the basis of all organic matter undergoes a constant cycle in nature by various heterotrophic bacteria. They process most of the energy flow and nutrients and act as a carbon sink. Mangrove leaves are leachable which, contains water soluble compounds, such as tannins and sugars (Cundell *et al.*, 1979), the remaining fraction of the organic matter consists of structural polymers as lignocelluloses which are directly used by the bacteria present in the soil and sediment of mangrove. Aerobic microorganisms have the enzymatic capacity for complete oxidation of organic carbon to CO_2, while the more important anaerobic degradation processes occur stepwise involving several competitive types of bacteria. Thus, large organic molecules are first split into small moieties by fermenting bacteria. These small molecules are then oxidized completely to CO_2 by anaerobic respiring bacteria using electron acceptors in the following sequence according to the energy yield: Mn^{4+}, NO_3^-, Fe^{3+} and SO^- (Canfield *et al.*, 2005). When all electron acceptors are exhausted and electron donors are in surplus, CH_4 is produced by fermentative disproportionation of low molecular compounds (e.g. acetate) or reduction of CO_2 by hydrogen or simple alcohols (Canfield *et al.*, 2005).

Litter retained within mangrove forests, either passively trapped by roots or handled by crabs, will eventually enter the microbial food chain, either in the form of uneaten remains buried in the sediment, fecal material or crab carcasses (Robertson, 1986) and contribute to the recycling of nutrients within the mangrove ecosystem. In general, crabs consume about half of the handled litter immediately, while the remainder is pulled into burrows to promote microbial colonization and leaching of tannins. Microbial decomposition of detritus in mangrove sediments is mostly mediated by bacteria, but marine mycelial decomposers belonging to Eumycotes (fungi) and Oomycotes (protoctista) are also important (Newell, 1996). The decay of mangrove litter in sediments begins with significant leaching of soluble organic substances. Newly fallen mangrove litter loses 20-40% of the organic carbon by leaching when submerged in seawater for 10-14 days (Camilleri and Ribi, 1986).

DOC released by leaching is either degraded within the sediment, in creeks or is exported by tides **(Fig.12)**. Some of the remaining particulate material is then rapidly hydrolyzed and solubilized by extracellular microbial enzymes and assimilated by these organisms. Most (80-90%) of the assimilated carbon is respired and lost as CO_2 to tidal water or emitted directly to the atmosphere (**Fig. 12**).

Methanogenic bacteria are inferior to anaerobic respiration processes in the competition for electron donors like hydrogen and acetate due to their rather low energy yield. Thus, a process like sulfate reduction can usually maintain concentrations of hydrogen and acetate at levels too low to fuel methanogens (Canfield *et al.*, 2005). Methanogenesis is therefore restricted to sediments where electron acceptors such as nitrate, metal oxides and sulfate are exhausted. As a consequence, methanogenesis is a dominating process in freshwater sediments, while sulfate reduction often dominates in marine sediments. Nevertheless, methanogenesis has been documented in many marine environments, particularly coastal marshes and mangrove swamps (Giani *et al.*, 1996, Kreuzwieser, 2003)

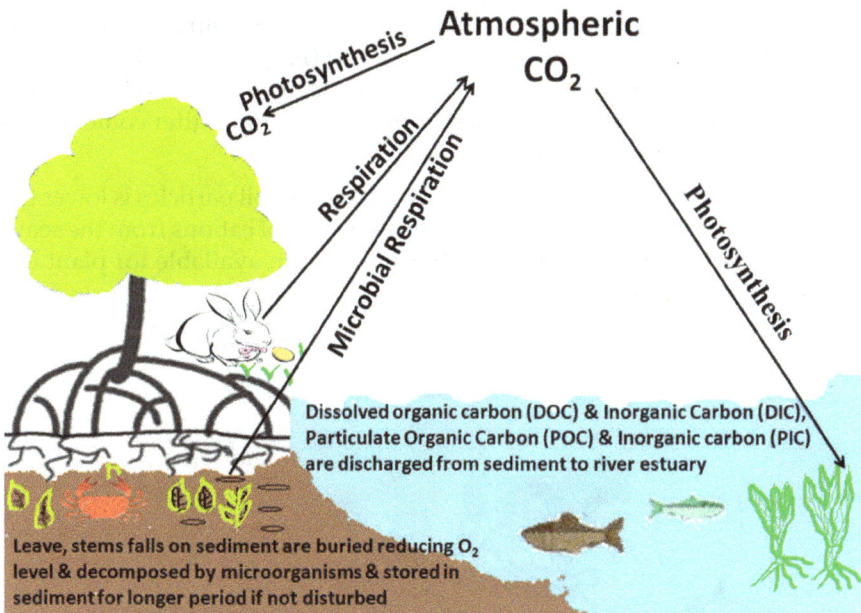

Fig.12: Carbon cycling in mangrove by microorganism

2.8 Nitrogen Content and its Cycling in Mangrove Soil

Nitrogen is cycled in mangrove environment by several bacteria **(Fig. 13)**. The ammonium present in sediments and derived from the degradation of nitrogenous organic compounds is probably converted to nitrate by nitrifying bacteria and is then assimilated by plants or anaerobically converted to N_2 by denitrification. The top layer of the soil and the thin layer of aerobic soil around the mangrove roots support populations of nitrifying bacteria that in turn can convert ammonium into nitrate for the plant; although nitrification rates are generally low (Shaiful *et al.*, 1986; Alongi *et al.*, 1992; Kristensen *et al.*, 1998). This process conserves the nitrogen within the ecosystem (Rivera-Monroy *et al.*, 1995a, b). However, high rates of denitrification have been found in mangrove ecosystems into which wastewater are discharged. High rates of denitrification deplete the nitrate and nitrite pools observed in mangrove soils (e.g. Alongi, 1994; Kristensen *et al.*, 2008). Fixation of molecular nitrogen is carried out intracellularly by various bacteria, e.g. *Azotobacter*, *Clostridium*, etc. As in other tropical forests (e.g. Cusack *et al.*, 2009), N_2 fixation in mangroves can be a significant source of N_2 (Holguin *et al.*, 2001). High levels of both light-dependent and light-independent N_2 fixation have been recorded in microbial communities living on the trees (Uchino *et al.*, 1984), in association with roots, in decaying leaves and on pneumatophores, as well as in the soil (Boto and Robertson, 1990). Benthic microbial mats are found in many intertidal mangrove habitats and can also contribute significantly to the N_2 cycle of the mangrove particularly when the mat is dominated by N_2-fixing cyanobacteria (Lee and Joye, 2006). Foliar uptake of N_2 in the form of ammonia from the atmosphere or from rainwater has also recently been suggested to be a potentially important source of N_2 for mangroves, particularly under conditions that favour ammonia volatilization (acidic, warm, flooded soils rich in organic matter) (Fogel *et al.*, 2008). Denitrifying bacteria are abundant in mangrove soils. Denitrification rates can be high due to the anaerobic conditions in combination with high organic matter content (Alongi, 1994; Corredor and Morell, 1994).

Furthermore, ammonium adsorption to mangrove soil particles is lower than in terrestrial environments due to the high concentration of cations from the seawater that compete for binding sites, making the ammonium available for plant uptake (Holmboe and Kristensen, 2002).

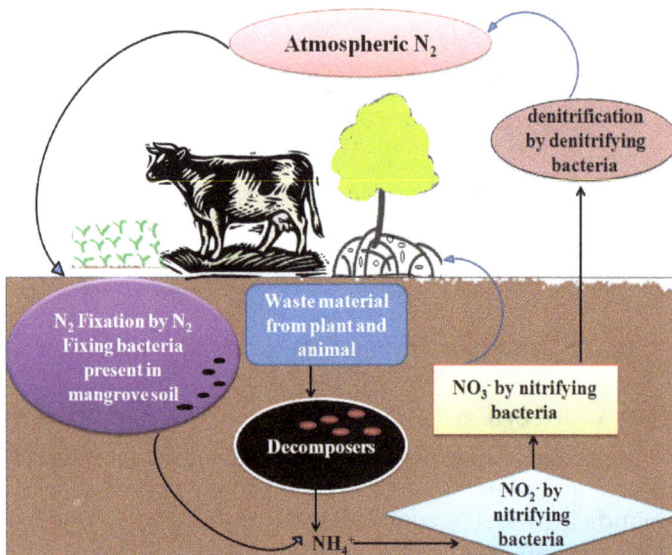

Fig. 13: Nitrogen cycling in mangrove by microorganism

2.9 Phosphorus Content and its Cycling in Mangrove Soil

Phosphorus found in mangrove soil as inorganic phosphate, are derived from weathering of parent rock. Organic phosphate in mangrove soils are produced by decayed plant, animal or microorganisms. Muddy mangrove soil has a strong capacity to absorb nitrates and insoluble phosphates carried out by the tides (Vazquez *et al.* 2000). Phosphorus usually precipitates in mangrove sediment due to its binding with various cations available in the interstitial water. As a result, phosphorus becomes largely unavailable to plants which is detrimental as phosphorous is vital to plant growth, especially in nutrient-limited mangrove environments. Phosphate- solubilizing bacteria, as potential suppliers of soluble forms of phosphorus, would have a great advantage for mangrove plants **(Fig. 14)**.

Mean concentrations of extractable phosphorus (extr.-P) across a mangrove forest gradient decrease with tidal height, it can become limiting in elevated areas (Boto and Wellington, 1983). Inorganic forms of phosphorus are represented in soil by primary minerals, such as apatite, hydroxyapatite and oxyapatite. The principal characteristic of these mineral forms is their insolubility. However, under appropriate conditions, they can be solubilized and become available for plants and microorganisms. Mineral phosphate can also be found associated with the surface of hydrated oxides of Fe, Al and Mn, which are poorly soluble and assimilable. A second major component of mangrove soil P is organic matter. Organic forms of P may constitute 30–50% of the total phosphorus in most soils, although it may range from as low as 5% to as high as 95% (Paul and Clark, 1988). Organic P in soils found largely in the form of inositol phosphate (soil phytate) which is synthesized by microorganisms and plants. It is the most stable organic forms of phosphorus found in soil, accounting for upto 50% of the total organic P (Dalal, 1977; Anderson, 1980; Harley and Smith, 1983). Other organic P compounds in soil are in the form of phosphor monoesters, phosphor diesters including phospholipids, nucleic acids, and phosphor triesters. Of the total organic phosphorus in soil, only approximately1% can be identified as nucleic acids or their derivatives (Paul and Clark, 1988). Many of these P compounds are high molecular-weight material which must first be bioconverted to either soluble ionic phosphate (Pi, HPO_4^{2-}, $H_2PO_4^-$), or low molecular-weight organic phosphate, to be assimilated by the cell (Goldstein, 1994).

Deficiency of soil P is one of the most important chemical factors restricting plant growth in soils. Phosphate solubilizing bacteria as potential suppliers of soluble phosphorus should confer a great advantage for plants through solubilisation and mineralization (Rodriguez and Fraga, 1999). How these bacteria solubilize phosphate is unclear, but culture experiments suggest that organic acids produced by the bacteria may dissolve calcium phosphate (Vazquez *et al.*, 2000). Their hydroxyl and carboxyl groups are able to form complexes with the iron and aluminium of corresponding phosphate compound in soil, thereby releasing bioavailable phosphate into the soil which can be utilized by plants (Gyaneshwar *et al.* 1998). Solubilization of phosphate-rich compounds is also carried out by the action of an enzyme called phosphatase. In all bacteria, this enzyme catalyzes the hydrolysis of a wide variety of phosphomonoesters and catalyzes a transphosphorylation reaction by transferring the phosphoryl group to alcohol in the presence of certain phosphate acceptors (Coleman, 1992). Bacteria solubilize phosphate in areas where the soil is oxygenated (e.g. near the mangrove roots) and may therefore, serve an important role in P uptake by the plant. Phosphate-solubilizing bacteria have been found in the roots of a number of mangrove species (Vazquez *et al.*, 2000; Rojas *et al.*, 2001; Kothamasi *et al.*, 2006) and their presence has been shown to increase the rates of other bacterial processes, such as nitrogen fixation (Rojas *et al.*, 2001).

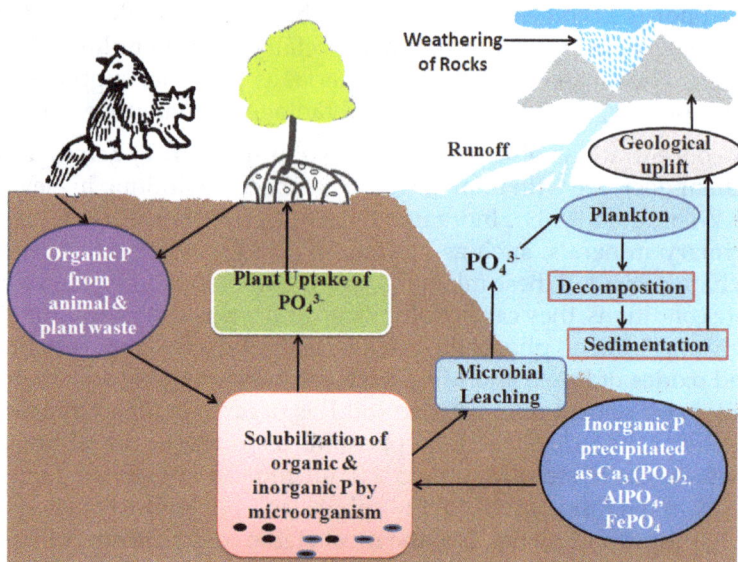

Fig. 14 Phosphorus cycling in mangrove by microorganisms

The uptake of soluble P by mangroves closely involves mutualistic inter relationships among bacteria, fungi, and tree roots. Arbuscular mycorrhizal fungi in the mangrove rhizosphere benefit from oxygen translocated by the trees to their roots and the presence of vesicles (nutrient storage organs) in the root cells of some mangrove species (Kothamasi *et al.*, 2006) suggests that fungal symbionts play a role in nutrient uptake. Phosphate solubilizing bacteria associated with the roots and fungi may release phosphate that could be taken up by the fungal hyphae and transferred to the host or taken up directly by the roots.

2.10 Sulphur Content and its Cycling in Mangrove Soil

Mangrove sediments are mainly anaerobic with an overlying thin aerobic sediment layer. Degradation of organic matter in the aerobic zone occurs principally through aerobic respiration whereas in the anaerobic layer decomposition occurs mainly through sulfate-reduction (Nedwell *et al.*, 1994; Sherman *et al.*, 1998). Aerobic and anaerobic microbial respiration, oxidize most of the organic carbon produced or deposited in mangrove sediments. Near the sediment surface, around crab burrows and along oxic root surfaces aerobic respiration occurs. The consumption of O_2 at these interfaces is usually so rapid that O_2 rarely penetrates more than 1-2 mm into the sediment (Kristensen *et al.*, 1998). A wide variety of anaerobic microorganisms mediate most carbon oxidation below the oxic zone of mangrove sediments. While aerobic respires consume litter and algal detritus deposited at or near the sediment surface, anaerobic respires are fuelled by detritus buried by accretion, by leaf-eating crabs and by below-ground root production in the form of dead biomass.

Microorganisms also play an important part in sulphur transformations in mangrove ecosystem **(Fig.15)**. Previous investigations also suggest that sulfate reduction may be an important pathway of organic matter mineralization in organic-rich deposits typical of mangrove forests (Alongi *et al.*, 1998). When sulphate is reduced by sulphate reducing bacteria, soluble sulphur compounds such as H_2S and HS are produced. These soluble sulphur compounds react with iron, reducing Fe (III) to Fe (II) and yielding pyrite (FeS_2). Consequently, most mangrove sediments contain high levels of reduced inorganic sulfur in the form of primarily pyrite (FeS_2) and elemental sulfur (So) and only negligible amounts of iron mono sulfides (FeS) (Holmer *et al.*, 1994). The close contact between oxygen and sulfide, on the other hand, also leads to extensive sulfide oxidation and subsequent acidification (Kristensen *et al.*, 1998). The acids generated will consume alkalinity and lower pH to values as low as 5 and prevent any storage of carbonates in mangrove sediment, even when the forests are surrounded by flats composed of carbonaceous coral sand.

Under anoxic conditions, sulphate-reducing bacteria reduce Fe to forms FeS_2 that are unfavourable for P binding (Holmer *et al.*, 1994), thereby releasing P to the pore water potentially for plant uptake. Hence, sulphate-reducing bacteria also play a pivotal role in increasing P availability in the soil (Sherman *et al.*, 1998). Sulfate-reducing bacteria can also contribute to the well-being of the ecosystem by fixing N_2 (Holguin *et al.*, 2001).

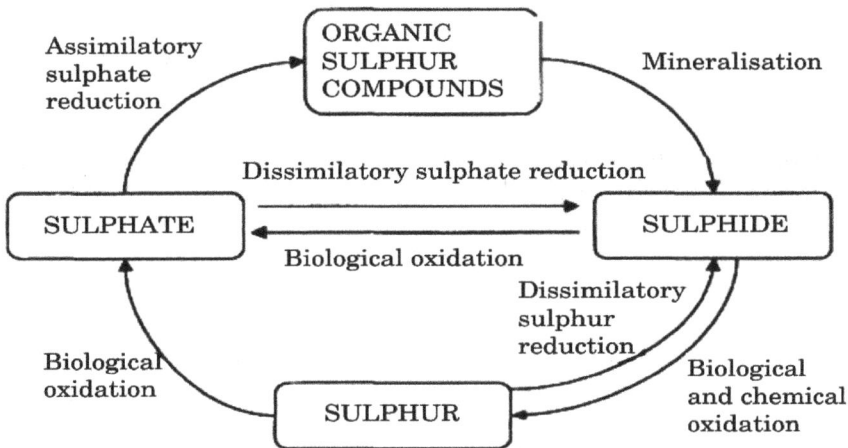

Fig. 15: The biological sulphur cycle (Janssen *et al.*, 1999)

Most of the degradation of organic matter occurs via sulphate reduction (Kristensen *et al.*, 1991). Oxidation of the soil around the roots can reverse the conversion of sulphate to sulphides, thus reducing the toxicity of the soil. However, this process also releases H^+ protons, which results in acidification of the soil. The high concentration of sulphate in seawater makes sulphide toxicity more probable in mangrove forests compared with terrestrial ecosystems (Raven and Scrimgeour, 1997).

2.11 Mangrove Vegetation of Different River Estuaries of Orissa Coast

Orissa has comparatively larger mangrove habitats in the country due to nutrient rich alluvial soils formed by different river deltas. However, major mangrove forests occur in the river Mahanadi delta and the Brahamani-Baitarani delta. The coast also has a smooth and gradual slope which provides larger areas for colonization of mangroves. As per an estimate the state has a total mangrove areas of 231 km² distributed over five coastal districts that include 82 km² very dense mangroves, moderately dense mangroves and 54 km² open mangroves cover in 2015 (State of Forest Report 2003). Mangroves of Bhitarkanika (Brahmani-Baitarani river deltas) are quite extensive and abundant as compared to those present in the Mahanadi delta. Apart from these, mangroves also occur in Devi river mouth, Budhabalanga and Subarnarekha river mouths are in highly degraded state. Mangroves occuring in the fringes if chilika lake have been disappeared.

Mangroves of river mouths of Subernarekha and Budhabalanga are at present in highly degraded state owing to heavy biotic pressure. Vegetation is represented by shrubby elements and stunted forms of tree species. The notable mangrove species are *Avicennia officinalis, Bruguiera gymnorrhiza, Sonneratia apetala, Excoecaria agallocha, Rhizophora mucronata, Bruguiera cylindrica, Ceriops decandra, Acanthus ilicifolius, Caesalpinia nuga, Myriostachya wightiana, Suaeda maritima, Porteresia coarctata,* etc. About 50 ha of degraded lands have been planted with seedlings of *Avecennia* sp., *Excoecaria agallocha, Ceriops decandra, Rhizophora mucronata, Bruguira gymnorrhiza, Aegiceras corniculatum,* etc. by the Odisha Forest Department during 2001-2002 (Panda *et al.*, 2013).

Bhitarkanika harbours luxuriant and wide-spread mangrove vegetation. The mangrove species occurring in Bhitarkanika exhibits two storey systems (Choudhury 1990). The ground flora is either poor or absent, top canopy is dominated by mangrove species (**Table-5**) viz., *Sonneratia apetala, Avicennia officinalis, A. alba, Excoecaria agallocha, Heritiera fomes, Pongamia pinnata,* while the second storey is composed of shrubby and under tree species such as *Brownlowia tersa, Kandelia candel, Lumnitzera racemosa, Rhizophora mucronata, Ceriops decandra, Cynometrairipa, Clerodendrum inerme, Aegiceras corniculatum, Hibiscus tiliaceus,* etc. Gregarious and luxuriant growth of *Avicennia spp.* and *Sonneratia apetala* are found along banks of river and creeks in Bhitarkanika. While *Pongamia pinnata* and *Xylocarpus granatum* occupy habitats close to the water bodies, in more elevated areas mixed forests of *Heritiera fomes* and *Excoecaria agallocha* are met with. While *Phoenix paludosa, Tamarix indica, Hibiscus tiliaceus, Heritiera littoralis* are found in pure formations, *Bruguiera gymnorrhiza, Cerbera manghas, Ceriops decandra,* etc. are found in low frequency.

The ground level, close to river banks and estuaries is muddy and studded with pneumatophores. Except a few patches of grasses herbaceous elements are devoided. However, a little distance from the river banks, *Acanthus ilicifolius, Acrostichum aureum, Flagellaria indica*, etc. in the moist areas and *Salicornia brackiata, Suaeda nudiflora, Tylophora indica*, etc. are in the dry region are usually found as herbaceous elements. The common associates as the category of climbers/twinners are *Derris trifoliata, Mucuna gigantean, Acanthus volubilis, Caesalpinia nuga, Dalbergia spinosa*, etc. Some mangrove species like *Cerbera manghas, Acanthus volubilies* and *Heritiera kanikensis* are found only in Bhitarkanika. Of these, the later two species are endemic to Odisha. Occurrence of 3 species each of *Avicennia, Heritiera, Sonneratia, Rhizophora* and *Xylocarpus* and 4 species of Bruguira are significant for Bhitarkanika forests (Panda *et al.*, 2013).

The dominant mangrove species occurring in Mahanadi delta are *Avicennia officinalis, A. marina, A. alba, Excoecaria agallocha, Rhizophora mucronata,* and *Sonneratia apetala*, etc. The mangroves in the region are degrading rapidly due to habitat alteration for agricultural system, development of prawn firms and of port facilities at Paradeep. These developmental activities moved many taxa towards threatened categories i.e., *Merope angulata, Tamarix dioica, Bruguira sexangula, Sonneratia caseolaris, Sonneratia alba, Sonneratia griffithii, Sarcolobus carinatus, S. globosus, Xylocarpus mekongensis* and *Dolichandrone spathacea* (**Fig.16**).

The estuary of Devi river is almost devoid of typical mangrove elements mainly due to habitat destruction connected with human settlement and paddy cultivation in the areas. Moreover, the ecological conditions have been changed due to formation of sand bars which considerably checked inundation. Only patches of *Acanthus ilicifolius, Tamarix troupii, Excoecaria agallocha, Myriostachya wightiana, Phoenix paludosa*, etc. are found in denuded condition. At places old stumps of *Avicennia sp.* and *Heritiera fomes* are reminiscent of the existence of past mangrove vegetation. In mud flats, *Suaeda maritima, Suaeda monoica, Sesuvium portulacastrum* and *Fimbristylis ferruginea* are common elements. State Forest Department and M.S. Swaminathan Research Foundation Chenai, have introduced a number of species i.e., *Avicennia officinalis, Sonneratia apetala, Aegiceras corniculatum, Ceriops decandra, Bruguira gymnorrhiza, Rhizophora apiculata, Avicennia marina*, etc. in the form of plantation in this area. In total, there are 12 true mangroves and 4 mangrove associates are present in Devi river mouth.

Mangroves of Chilika Lake: The typical mangroves have disappeared from the fringes of the Chilika lake and its adjoining regions. Only mangrove associates and transitory taxa like *Aegiceras corniculatum, Azima tetracantha, Salvadora persica, Cressa cretica*, etc. are found at places. However, species like *Clerodendrum inerme, Excoecaria agallocha*, etc. as reported by Narayan swami and Cater (1922) could not be traceable at present. About 10 ha of land near Sipakuda and adjoining area have been planted with species i.e., *Rhizophora apiculata, Avicennia marina, A. alba, A. officinalis* and *Kandelia candel* by Forest Department of Odisha (**Table-5**).

Table 5: A comparison of true mangrove species in Orissa

Name of the mangrove	Name of the species	Reference
Subernarekha and Budhabalanga river mouths	*Avicennia officinalis, Bruguiera gymnorrhiza, Sonneratia apetala, Excoecaria agallocha, Rhizophora mucronata, Bruguiera cylindrica, Ceriops decandra, Acanthus ilicifolius, Caesalpinia nuga, Myriostachya wightiana, Suaeda maritima, Porteresia coarctata,*	Thatoi and Biswal, 2008
Bhitarkanika (Brahmani and Baitarani river delta)	*Sonneratia apetala, Avicennia officinalis, A. alba, Excoecaria agallocha, Heritiera fomes, Pongamia pinnata, Brownlowia tersa, Kandelia candel, Lumnitzera racemosa, Rhizophora mucronata, Ceriops decandra, Cynometrairipa, Clerodendrum inerme, Aegiceras corniculatum, Hibiscus tiliaceus, Salicornia brackiata, Suaeda nudiflora, Tylophora indica, Salicornia brackiata, Suaeda nudiflora, Tylophora indica, Salicornia brackiata, Suaeda nudiflora, Tylophora indica, Salicornia brackiata, Suaeda nudiflora, Tylophora indica, Derris trifoliata, Mucuna gigantean, Acanthus volubilis, Caesalpinia nuga, Dalbergia spinosa, Cerbera manghas, Acanthus volubilies* and *Heritiera kanikensis*	Thatoi and Biswal, 2008
Mahanadi delta	*Avicennia officinalis, A. marina, A. alba, Excoecaria agallocha, Rhizophora mucronata, and Sonneratia apetala, Merope angulata, Tamarix dioica, Bruguira sexangula, Sonneratia caseolaris, Sonneratia alba, Sonneratia griffithii, Sarcolobus carinatus, S. globosus, Xylocarpus mekongensis* and *Dolichandrone spathacea.*	Thatoi and Biswal, 2008
Devi river mouth	*Acanthus ilicifolius, Tamarix troupii, Excoecaria agallocha, Myriostachya wightiana, Phoenix paludosa, Heritiera fome, Suaeda maritima, Suaeda monoica, Sesuvium portulacastrum* and *Fimbristylis ferruginea, Avicennia officinalis, Sonneratia apetala, Aegiceras corniculatum, Ceriops decandra, Bruguira gymnorrhiza, Rhizophora apiculata and Avicennia marina*	Thatoi and Biswal, 2008
Chilika lake	*Aegiceras corniculatum, Azima tetracantha, Salvadora persica, Cressa cretica, Clerodendrum inerme, Excoecaria agallocha, Rhizophora apiculata, Avicennia marina, A. alba, A. officinalis* and *Kandelia candel (Planted)*	Thatoi and Biswal, 2008

Xylocarpus granatum   Kandelia candel

Lumnitzera racemosa   Sonneratia grifithii

Sesuvium
portulacastrum

Pnematophores of
officinalis

Fig. 16. Mangrove species found in Odisha coast (Thatoi & Biswal, 2008)

2.12 Physico-chemical Characteristics of Water from Mangroves of Bhitarkanika and Mahanadi Delta.

Physico-chemical analysis of water samples from the mangrove forests of Bhitarkanika and Mahanadi delta were carried out by the authors. The physico-chemical parameters of water quality includes (pH, temperature, conductivity, TDS, DO, calcium, magnesium, total hardness, nitrate, phosphate and chloride (**Table-6**). It was found that there is little variation among the temperature, pH, dissolved oxygen, chloride, phosphate, calcium, magnesium and total hardness of both the mangrove waters whereas the conductivity and amount of nitrate content varied widely. In this study, water temperature among the different study sites of mangroves of Mahanadi delta showed little variation (24.2 °C – 30.9 °C). It has been observed that the variation of temperature of water among the study sites supports the finding of Mishra et al. (2008) who has reported the same trend of water temperature in the mangrove forest of Bhitarkanika, Odisha. pH of water at all the study sites showed a narrow range of variation. Water samples collected from different sites of mangroves of Bhitarkanika showed variation in pH from 6.7 to 7.9. Similarly the pH of the water samples collected from mangroves of Mahanadi delta varies from 6.05 to 8.6. The observed value of pH of water samples collected from both the mangroves indicates that the water is slightly acidic to merely alkaline in nature. The conductivity of water samples of mangroves of Mohanadi delta at different study sites was found variable (5.16 -17.3 mS/cm). The TDS value at all the study sites in both the mangroves did not vary much. The total amount of dissolved oxygen (DO) among the different sites in mangroves of Mahanadi delta ranged from 2.9 mg/L to 10.9 mg/L. The similar trend is also observed in mangroves of Bhitarkanika. The DO content of water samples collected from mangroves of Bhitarkanika varied from 3.7-11.36 mg/L. Chloride content of water samples collected from mangroves of Mahanadi delta was observed in the range of 4389- 9575.21 mg/L. The chloride content of water samples collected from mangroves of Bhitarkanika varied from 9118.70-10,937.61 mg/L. Generally, estuarine mangrove waters have low content of dissolved inorganic phosphorous and nitrogen. There is a significant variation of phosphate and nitrate content was observed during the sampling period. On an average, the phosphate content of water samples collected from mangroves of Mahanadi delta ranges from 0.55 mg/L to 2.59 mg/L throughout the study period. There is a significant variation in calcium content in the water samples collected from mangroves of Mahanadi delta during the investigation which varied from 125.4 mg/L to 400.8 mg/L. In comparison to Mahanadi mangrove water, the water samples collected from mangroves of Bhitarkanika, showed lower calcium content i.e. 233-281.23 mg/L. The magnesium content of water samples collected from both the mangrove forests did not show much variation. In an average, the magnesium content of water samples collected from both the mangroves varied from 119.5-474.13 mg/L.

Table 6: Water physico-chemical characteristics of mangroves of Bhitarkanika and Mahanadi delta. (Data are average values of different study sites)

physicochemical characteristics	Bhitarkanika Mangrove	Mahanadi Mangrove
Water Temperature (°C)	21.20 – 35.80	24.2°C – 30.9°C
pH	6.70 – 7.90	6.05-8.6
Conductivity (mS/cm)	3.25 – 8.45	5.16 -17.3
Total Dissolved Solid (((mg/L)	7.45 – 11.23	4.51 – 11.90
Dissolved Oxygen (mg/L)	3.7- 11.36	2.9 - 10.9
Chloride (mg/L)	9118.70 -10937.61	4389.25 – 9575.21
Phosphate (mg/L)	0.14 -1.08	0.55 - 2.59
Nitrate (mg/L)	3.16 -3.68	13.03 - 24.01
Calcium (mg/L)	233 - 281.23	125.4 - 400.8
Magnesium (mg/L)	119.5-380.45	153.16 - 474.13
Total Hardness (mg/L)	**1136.16 - 1769.42**	**800.22 – 2090.10**

2.13 Soil physico-chemical and Nutrient Contents of Mangroves of Bhitarkanika

Soil, pH, electrical conductance (Eh) and nutrient contents (total nitrogen, phosphorous, carbon and potash) of different sites (Rangani, Mahisamunda, Habalaganda, Dangamal and Kalibhanjadian) from Bhitarkanika magrove forests during four different seasons (Rainy, Autumn, Winter, Summer) were given (Table 7 and Fig. 17). The soil physico-chemical characters such as pH, salinity and total N, P, K and C contents did not follow any common seasonal and spacial trend. The pH of the different sites was limited within a narrow range of 6.02-7.89 which was acidic (6.0-6.6) during the winter but neutral to marginally alkaline in the other seasons. The redox potential of the soils was comparable in different seasons but in the summer it almost doubled. Total N content of the five sites ranged between 200.6-285.5 kg/ha and did not vary significantly in any given season. However, it gradually declined from the rainy season to the summer season. The trend of seasonal N content was followed by P contents also but unlike the N, they were variable between 9.0-24.0 kg/ha among the sites. The P level was optimum (24.00 kg/ha) at site 4 (Dangamal) and minimum (9.80 kg/ha) at site 1 (Rangani). The K level was variable within 1053-2378 kg/ha among the sites and it was elevated 2-5 times in the rainy season. Total carbon (C) content of the sites (0.11-0.59%) did not follow any common seasonal trend, which was optimum in the winter and minimum in the autumn. It was more at site 5 (Kalibhanjadian). Nevertheless, overall soil nutrition status was superior in the winter followed by the rainy, summer and autumn seasons.

2.14 Soil physico-chemical and Nutrient Contents in Mangroves of Mahanadi Delta

The spatial distribution of soil characteristics such as pH, salinity (E.C.), total N, P, K and total carbon of mangroves of Mahanadi delta were analyzed from six different sites and their average seasonal values were shown in Fig. 18 a-f and Table-8. The pH of different sites was limited within a narrow range of 5.5-6.8 which were slightly acidic in nature. Higher pH value (6.8) was observed at site 3 (Triveni) during winter season. The E.C. of the soil samples fluctuated at the different sites of the mangrove environment with mean values between 0.5 - 4.9 mS/cm. Maximum E.C. was recorded at site 5 (Atharbanki) during the summer season and the minimum E.C. value was observed at site 3 and site 4 during rainy season. Total nitrogen content of the six sites ranged between 0.037- 0.094% with maximum value in winter season (0.094%). However, there is a marginal variation of nitrogen content was observed in all the sites during the study period. Phosphorus of the soil exhibits wide variation from 8.0-138.0 kg.ha^{-1} among the study sites with a maximum in rainy season. Significantly higher phosphorous content was observed in soils of IFFCO and Atharbanki (near PPL) sites. The potassium level was variable within 289-498 kg.ha^{-1} among the study sites. Total organic carbon among the study sites varied between 0.38- 0.99% showing maximum in summer and minimum during winter.

Table 7: Soil physico-chemical characteristics of mangroves of Bhitarkanika in four different seasons

Season	Sites	pH	Salinity	Nitrogen	Phosphorus	Potash	Carbon
Rainy	Rangani	7.25 ± 0.05	7.60 ± 0.0	285.50 ± 1.04	12.00 ± 0.98	1863.0 ± 1.00	0.25 ± 0.05
	MaMahisamunda	7.89 ± 0.25	7.20 ± 0.25	279.08 ± 1.00	18.00 ± 0.98	1663.0 ± 1.02	0.32 ± 0.08
	Habalaganda	7.22 ± 0.54	7.30 ± 0.65	245.25 ± 1.06	15.00 ± 0.89	1075.0 ± 1.00	0.41 ± 0.07
	Dangamal	7.75 ± 0.45	6.40 ± 0.35	276.18 ± 1.44	21.00 ± 0.99	2363.0 ± 0.09	0.22 ± 0.58
	Kalibhanjadian	7.47 ± 0.54	8.00 ± 0.76	280.43 ± 1.34	15.00 ± 0.40	1988.0 ± 1.89	0.53 ± 0.75
Autumn	Rangani	7.32 ± 0.85	8.50 ± 0.99	277.80 ± 1.05	11.45 ± 0.65	1563 ± 1.0	0.41 ± 0.76
	MaMahisamunda	7.56 ± 0.66	8.60 ± 0.78	268.50 ± 1.60	12.80 ± 0.56	1638 ± 1.23	0.18 ± 0.34
	Habalaganda	7.62 ± 0.78	7.70 ± 0.87	237.25 ± 1.00	13.54 ± 0.65	1053 ± 0.76	0.36 ± 0.32
	Dangamal	7.05 ± 0.89	8.80 ± 0.89	265.80 ± 1.32	19.80 ± 0.78	2168 ± 0.59	0.44 ± 0.45
	Kalibhanjadian	7.19 ± 0.45	9.50 ± 0.86	213.0 ± 1.54	9.00 ± 0.56	1898.0 ± 1.2	0.25 ± 0.32
Winter	Rangani	6.59 ± 0.66	8.90 ± 0.98	259.50 ± 1.00	15.00 ± 0.87	1563 ± 1.20	0.48 ± 0.01
	MaMahisamunda 6	6.62 ± 0.76	8.90 ± 0.76	247.50 ± 1.05	15.00 ± 0.76	1675 ± 1.57	0.39 ± 0.23
	Habalaganda	6.65 ± 0.85	8.00 ± 0.54	230.05 ± 1.32	12.00 ± 0.54	1025 ± 1.00	0.49 ± 0.45
	Dangamal	6.02 ± 0.78	9.00 ± 0.76	245.60 ± 1.05	24.00 ± 0.99	1988 ± 1.32	0.44 ± 0.34
	KKKalibhanjadian	6.30 ± 0.67	9.80 ± 0.79	204.045 ± 1.04	13.00 ± 0.89	2015 ± 1.23	0.59 ± 0.43
Summer	Rangani	7.42 ± 0.43	18.40 ± 0.87	217.40 ± 1.00	9.80 ± 0.76	1878 ± 1.20	0.39± 0.40
	MaMahisamunda 6	7.40 ± 0.54	17.69 ± 0.45	211.05 ± 1.23	11.07 ± 0.24	1685 ± 1.23	0.11 ± 0.34
	Habalaganda	7..36 ± 0.23	19.00 ± 0.58	209.50 ± 1.45	8.50 ± 0.05	1095 ± 1.03	0.26 ± 0.23
	Dangamal	7.01 ± 0.76	16.50 ± 0.45	216.60 ± 1.54	19.80 ± 0.08	2378 ± 1.04	0.48± 0.34
	Kalibhanjadian	7.29 ± 0.58	19.5 ± 0.34	198.80 ± 1.34	9.0 ± 0.08	2110 ± 1.02	0.20 ± 0.43

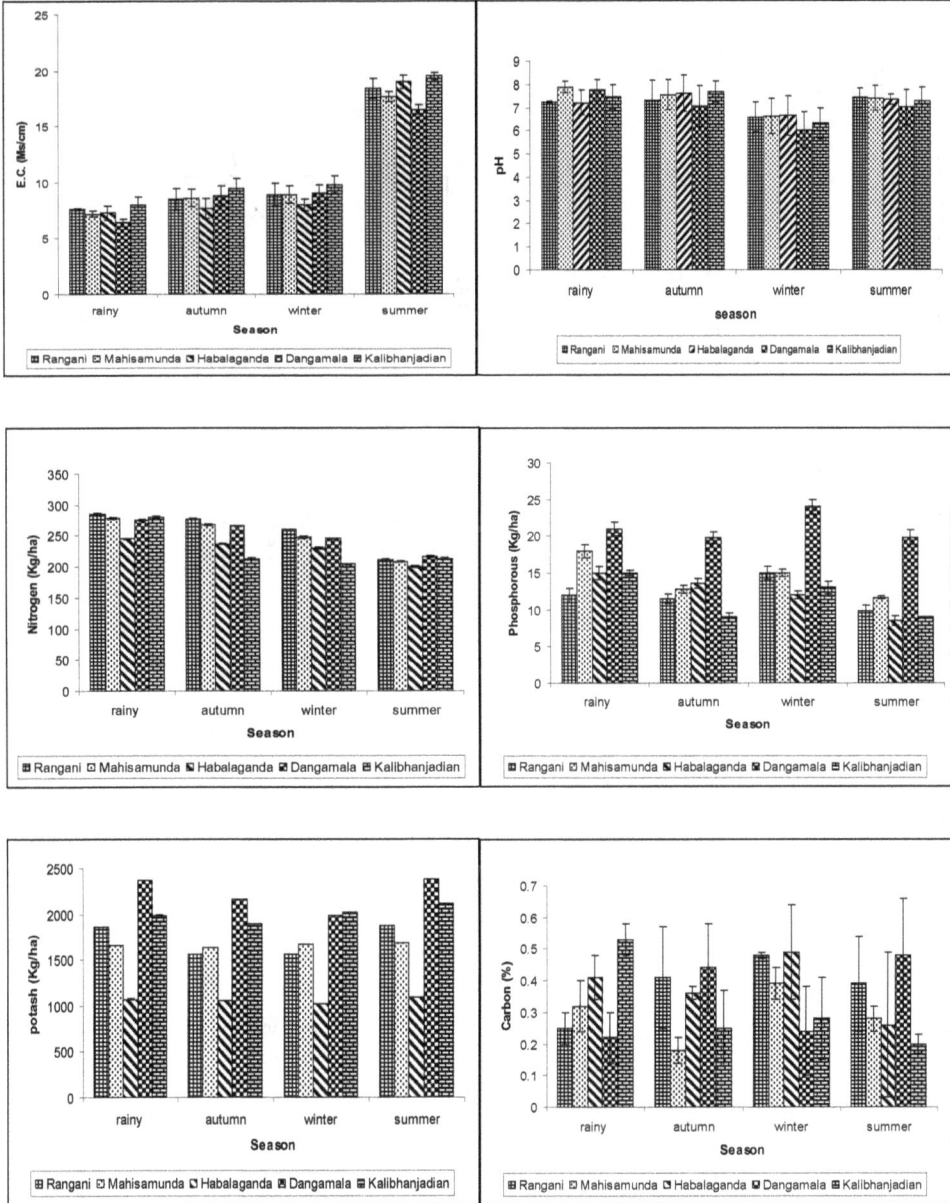

Figure 17: Seasonal variation nutrient contents content of different soil samples collected from mangrove sites of Bhitarkanika, Odisha.

Table 8: Soil physico-chemical characteristics of mangroves of Mahanadi river delta in four different seasons

Season	Sites	pH	Salinity	Nitrogen	Phosphorus	Potash	Carbon
Rainy	Jambu	6.20±0.05	0.80±0.17	0.05±0.01	12.00±0.57	410.00±1.99	0.67±0.01
	Kharsi	5.70±0.05	0.73±0.17	0.06±001	26.00±1.73	392.10±1.00	0.58±0.05
	Triveni	5.60±0.15	0.50±0.05	0.05±0.01	23.00±1.73	408.00±0.57	0.48±0.02
	Nuagada	5.50±0.15	0.50±0.00	0.04±0.01	25.00±1.52	289.00±2.59	0.42±0.01
	Atharbanki	6.50±0.10	1.33±0.01	0.06±0.01	138.00±4.04	302.00±1.00	0.45±0.01
	IFFCO	6.70±0.15	1.31±0.05	0.06±0.01	102.00±5.77	400.00±1.73	0.57±0.01
Autumn	Jambu	6.40±0.15	1.93±0.01	0.07±0.01	8.00±1.15	496.00±1.73	0.56±0.01
	Kharnasi	6.20±0.05	1.20±0.17	0.06±0.00	19.00±1.00	498.00±4.61	0.65±0.01
	Triveni	5.70±0.11	1.60±0.15	0.06±0.00	19.00±1.20	444.00±0.57	0.50±0.01
	Nuagada	5.50±0.15	1.20±0.11	0.05±0.00	20.00±6.42	370.00±2.64	0.38±0.01
	Atharbanki	6.40±0.10	1.59±0.01	0.06±0.00	123.00±2.07	460.00±4.93	0.49±0.05
	IFFCO	6.60±0.10	1.28±0.01	0.08±0.00	88.00±1.15	468.00±0.57	0.78±0.01
Winter	Jambu	5.90±0.16	2.05±0.01	0.07±0.01	9.00±1.00	460.10±5.70	0.79±0.01
	Kharnasi	6.70±0.10	2.94±0.58	0.08±0.00	22.00±1.99	488.00±10.00	0.82±0.01
	Triveni	6.81±0.05	2.60±0.35	0.07±0.01	14.00±1.99	466.00±0.57	0.51±0.00
	Nuagada	6.70±0.15	2.50±0.58	0.06±0.00	17.00±2.30	430.00±5.77	0.41±0.01
	Atharbanki	6.28±0.05	3.57±0.56	0.09±0.00	105.00±5.77	470.00±4.04	0.54±0.01
	IFFCO	6.53±0.05	3.31±0.01	0.09±0.01	70.00±2.88	480.00±5.77	0.73±0.01
Summer	Jambu	6.63±0.11	1.93±0.01	0.07±0.00	9.60±0.11	450.00±5.27	0.78±0.01
	Kharnasi	6.30±0.57	4.08±0.00	0.07±0.00	16.00±1.00	460.00±1.54	0.99±0.05
	Triveni	6.20±0.19	3.02±0.01	0.06±0.01	15.00±1.00	312.00±1.52	0.62±0.01
	Nuagada	6.40±0.19	3.10±0.11	0.04±0.00	16.30±0.09	310.00±2.07	0.47±0.01
	Atharbanki	6.10±0.05	4.90±0.17	0.07±0.00	103.00±5.77	385.00±5.77	0.89±0.01
	IFFCO	6.00±0.00	4.75±0.02	0.07±0.00	65.00±3.46	455.00±2.30	0.87±0.02

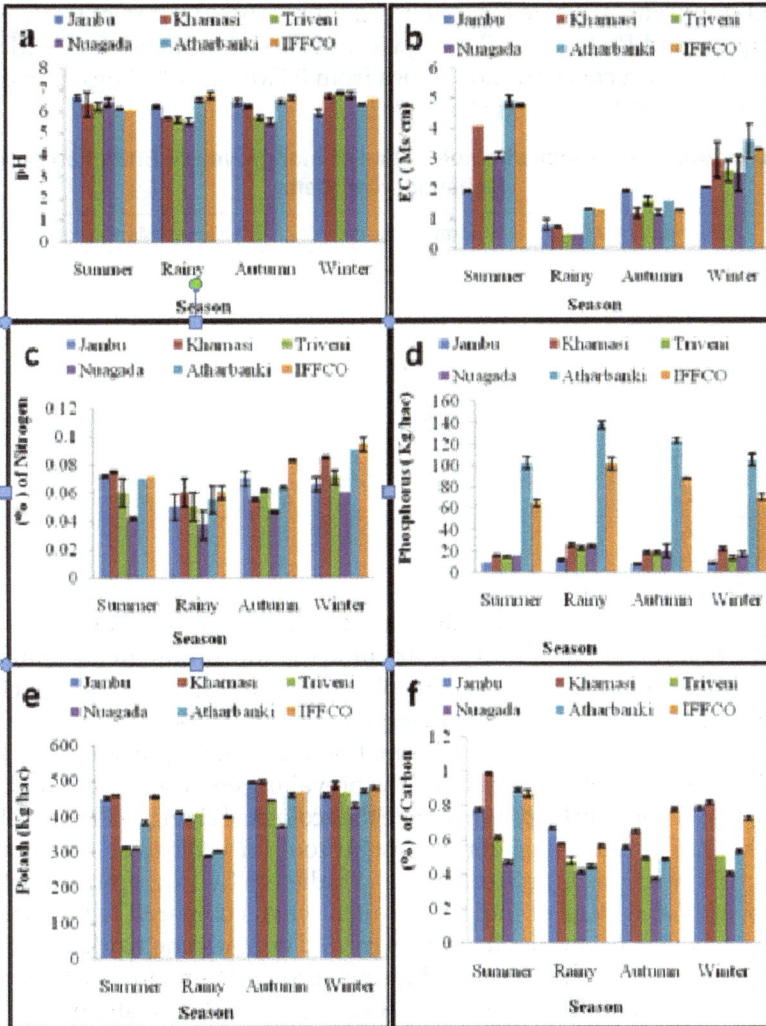

Fig. 18: Physico-chemical characteristics of mangroves soils of Mahanadi delta (a) pH, (b) Salinity, (c) Nitrogen, (d) Phosphorus, (e) Potash, (f) Carbon

2.15 Soil Heavy Metal Contents from Mangroves of Mahanadi Delta and Bhitarkanika

The average concentrations (µg/g) of various heavy metals *viz*. Fe, Cu, Ni, Cd, Pb, Zn and Co in the soil samples at locations of the mangrove environment were observed in the range of 14810-63370 µg/g, 2.8-32.6 µg/g, 16.4-55.7 µg/g, 1.8-7.9 µg/g, 14.3-54.7 µg/g, 24.4-132.5 µg/g and 13.3-48.2 µg/g respectively **(Table 9)**. The normal value of Cu in soil is 30 µg/g and the observed value of Cu (2.8-32.6 µg/g) are above the prescribed limit. The average concentration of Ni in the soil samples were found below the permissible limit of (IS) (75-150 µg/g) with maximum (55.7 µg/g) and minimum (13.4 µg/g). The maximum concentration of Cd (7.9 µg/g), Co (48.2 µg/g) and Pb (14.3-54.7) µg/g were observed at mangrove of Mahanadi delta whereas the Zn concentration varies from 0.7Mg/g (132.5 µg/g) in different sites (Behera et al. 2013; Sarangi et al., 2002).

Table 9: Soil heavy metal characteristics of mangroves of Bhitarkanika and Mahanadi river delta

Soil heavy metal	Bhitarkanika angrove	Mahanadi Mangrove
Fe (µg/g)	32-41	14810-63370
Cu (µg/g)	2.6-2.9	2.8-32.6
Ni (µg/g)	ND	16.4-55.7
Cd (µg/g)	ND	1.8-7.9
Pb (µg/g)	ND	14.3-54.7
Zn(µg/g)	0.7-1.2	24.4-132.5
Co (µg/g)	ND	13.3-48.2

Conclusion

Mangrove soil and water remain important area of research because of their ecological significance. Most of the nutrient cycle and productivity of mangrove depends on soil and water quality. The physico chemical parameters such as the temperature, pH, salinity, dissolved oxygen and nutrients are the major master factors of costal water ecosystem. Information on various physico-chemical and biological process, which are controlling the prevailing environmental conditions of the region, will eventually helps to evaluate the ecological changes. Physico-chemical characteristics of water would form a useful tool for ecological assessment and monitoring of coastal mangrove ecosystem. Studies on hydrography of backwaters of mangroves of east coast of India are limited when compared to the mangroves of west coast. However, studies have been carried out by some workers in different locations along the east coast which include mangrove of Sundarbans (Rahman et al. 2013), Bhitarkanika and Mahanadi delta (Chauhan and Ramanathan 2008; Mishra et al; 2009 and Behera et al. 2012; Pradhan et al., 2014), Krishna and Godavari delta (Satyanaranyan and Krishna 2016) and Pichavaram (Ravichelvan et al 2015) in the east coast of India.

Increased anthropological activities and increase in organic and inorganic pollutants leads to deteoration of the mangrove wetlands. Changes in physico chemical parameters are serious problems of aquatic organisms and affect the distribution of biotic elements of the mangrove environment. Increased discharge of effluents from industries, municipality drainages and aquaculture ponds contribute to the pollution of the mangrove ecosystem. Sediment is pivotal components of aquatic ecosystems where important transformations and exchange processes take place (Levine *et al.*, 2001). Mangrove sediment of Bhitarkanika is highly silty, clayey and loamy which are about (3-4) m in depth. The soil is saline and sustained by salt tolerant mangrove vegetation, typical to this environment. The mangrove area is inundated by the tidal water twice a day. This is a dynamic ecosystem which shows diurnal and seasonal variation of salinity, temperature, pH and tidal inundation.

Similarly mangrove sediment of Mahanadi delta is silty, clayey and sandy clayey in varying proportion. The silty and clayey soils are chiefly derived from land are nutrient rich. The organic matter contents in soil varies from 0.01% - 2.02% (Mahalik, 2000). The mangrove area like Bhitarkanika mangrove is inundated by the tidal water twice a day. Apart from this, the soil physico-chemical and nutritional status also varies to a great extent within study sites and seasons which have been assessed in the present study. The pH and salinity of the soil generally varies between 5.5-6.8 and 0.5 - 4.9 (mS/cm) respectively. Similarly soil N, P, K. and organic C values ranges between 0.037-0. 094 %, 8.0-138.0 kg.ha^{-1}, 289-498 kg.ha^{-1} (kg/ha) and 0.38- 0.99% respectively. The soil nutrient content in mangrove environment is also very high due to the decomposition of mangrove leaf litter and other plant materials. Analysis of soil showed high microbial populations related to major nutrient recycling such as C, N, S and P etc. which might play an important role for transformation of organic matter and nutrient recycling in mangrove ecosystem. So far nutrition is concerned, mangroves of Odisha has not fully explored so far. Few studies have been undertaken till date. Except Bhitarkanika and Mahanadi delta, other mangroves of Odisha are yet to be studied. Besides heavymetal pollution of mangrove ecosystem of Odisha has not been studied much. The scientific study of all the mangrove habitats will give a clear picture about the ecology of the areas.

MICROBIAL DIVERSITY OF MANGROVE ECOSYSTEM

Microbial Diversity of Mangrove Ecosystem

Both the soil and sediment probably represent some of the most complex microbial habitats on the earth. One gram of soil may contain several thousand species of bacteria (Torsvik *et al.*, 1990). To understand nature and diversity of these microorganisms it is essential to isolate and identify them. Methods to study the diversity of microbial populations in soil and water can be categorized into groups i.e. traditional biochemical techniques and molecular techniques. The traditional biochemical methods are of high significance in the study of microbial diversity. The most traditional method for assessment of microbial diversity is selective and differential plating and subsequent viable counts. Being fast and inexpensive, these methods provide information about active and culturable heterotrophic segment of the microbial population. Subsequent identification of microorganisms is based on their phenotypic characteristics. Traditional methods for characterising microbial communities have been based on the analysis of the culturable portion of the bacteria. Due to non-culturability of the major fraction of bacteria from natural microbial communities, the overall structure of the community has been difficult to interpret. Recent studies to characterise microbial diversity have focussed on the use of methods that do not require cultivation, yet provide measures based on genetic diversity. During the past decade, the development of molecular techniques using nucleic acids has led to many new findings in the studies of microbial diversity (Amann *et al.*, 1995). As a basic approach to clarify the microbial communities, 16S rRNA genes are amplified by PCR from nucleic acids extracted from environmental samples and then the PCR products are cloned and sequenced. This approach can avoid the limitation of the traditional culturing techniques for assessing the microbial diversity in their natural environments.

Mangrove soil is rich in organic matter due to high litter deposition. They provide a unique ecological site to different microorganisms. Microbial activity is responsible for major nutrient transformations within a mangrove ecosystem (Alongi *et al.*, 1993). The microbial community in the mangrove sediment is strongly influenced by bio-geographical, anthropological and ecological properties which include food web in the ecosystem, nutrient cycling and the presence of organic and inorganic matters. Mangrove ecosystem shows diversity of microorganisms. All microbial forms such as bacteria, fungi, cyanobacteria, microalgae, macroalgae, fungus like protists and actinomycetes have been reported in this ecosystem. In tropical mangroves, bacteria and fungi constitute 91% of the total microbial biomass, whereas algae and protozoa represent only 7% and 2%, respectively (Alongi, 1988). Due to high litter deposition and low rates of organic matter decomposition, mangrove sediment is anoxic soils (Nedwell *et al.,* 1994; Komiyama *et al.,* 2008). The degradation of organic matter in mangrove sediments is mediated by various microbial fermentation and respiration processes. Among these microorganisms, bacteria are responsible for most of the carbon flux, the energy flow and nutrients recycling of the mangrove ecosystem. The common bacterial groups found in the mangroves are carbon transforming bacteria, sulphate-reducing bacteria, N_2 fixing bacteria, phosphate solubilizing bacteria, photosynthetic anoxygenic bacteria, methanogenic bacteria etc.

Decomposition of organic carbon present in the mangrove sediments are mediated by different bacterial communities. Most of the heterotrophic communities of surrounding estuarine and marine ecosystems have a pivotal role in the mangrove carbon cycle (Odum and Heald, 1975a). Roughly 30 to 50% of the organic matters in mangrove leaves are leachable; containing water soluble compounds, such as tannins and sugars (Cundell *et al.,* 1979). The remaining fraction of the organic matter consists of structural polymers as lignocelluloses which are directly used by both the aerobic and anaerobic the micoroorganisms present in the soil and sediment of mangrove. Aerobic microorganisms have the enzymatic capacity for complete oxidation of organic carbon to CO_2, while the more important anaerobic degradation processes occur stepwise involving several competitive types of bacteria. The, large organic molecules are first split into small moieties by fermenting bacteria. These small molecules are then oxidized completely to CO_2 by anaerobic respiring bacteria using electron acceptors in the following sequence according to the energy yield: Mn^{4+}, NO^{3-}, Fe^{3+} and SO^- (Canfield *et al.,* 2005). When all electron acceptors are exhausted and electron donors are in surplus, CH_4 is produced by fermentative disproportionation of low molecular compounds (e.g. acetate) or reduction of CO_2 by hydrogen or simple alcohols (Canfield *et al.,* 2005). The methane thus produced is again utilized by methanogenic bacteria (*Methanoccoides methylutens* sp., etc.). In mangrove sediment, the primary factors controlling microbial carbon transformations are vegetation type, root activity, infaunal activity, sediment type, organic input, tidal elevation, salinity and eutrophication (Robertson *et al.,* 1992). It has been found that, root activity (leaching), infaunal activity (crabs) and organic input (quantity and quality) are important for CO_2 generation via respiring bacteria, while salinity (sulfate) is among the key factors controlling the importance of CH_4 generation (Lu *et al.,* 1999). Photosynthetic bacteria also play an important role in the production of organic matter in mangroves as well as purification of the polluted water of the ecosystem. Photosynthetic anoxygenic bacteria (*Chloronema, Chromatium, Beggiatoa, Thiopedia, Leucothiobacteria* sp.) have the ability to grow in the anaerobic environment like mangrove.

Mangrove soils are anaerobic environment and rich in sulphate and organic matter. Other than organic carbon transforming bacteria, sulfate reducing bacterial communities are also present in mangrove water and soil such as *Desulfovibrio, Desulfotomaculum, Desulfosarcina, Desulfococcus* sp. etc.). These bacterial communities get their energy by reducing elemental sulfur to hydrogen sulfide. Nitrogen fixation was considered as the major source of combined nitrogen input in mangrove forest habitat (Hicks and Silvester, 1985). High rates of nitrogen fixation have been found associated with dead and decomposing leaves, pneumatophres, tree bark, cyanobacterial mats covering the surface of the sediment (Zuberer and Silver, 1978). Several N₂ fixing bacteria such as *Azospirillum, Azotobacter, Rhizobium, Clostridium, Klebsiella* sp., etc. have been reported in mangroves. In mangrove environment, phosphorus usually precipitates due to the abundance of cations in the interstitial waters surrounding mangrove sediment, making it non-bioavailable to plants and most other organisms. Phosphate solubilising bacteria (*Bacillus, Paenibacillus, Xanthobacter, Vibrio, Proteolyticus, Enterobacter, Kluyvera, Chryseomonas* and *Pseudomonas* sp., etc.) play a major role in converting phosphate into a bioavailable form which can be directly taken up by plants.

Besides, various groups of fungi such as ligninolytic, cellulolytic, pectinolytic, amylolytic and proteolytic as well as actinomycetes are present in mangrove ecosystem (Kathiresan and Bingham, 2001). Among the algae, groups like *Chlorophyta, Chrysophyta, Phaephyta, Rhodophyta* and *Cyanophyta* are dominant in the mangrove ecosystem (Sen and Naskar, 2003). Several studies have shown the uniqueness of mangrove sediments with respect to their microbial composition (Urakawa *et al.*, 1999). Studies on microbial diversity in the mangrove sediments are important to understand the process of biogeochemical cycling and pollutants removal (Roy *et al.*, 2002). Moreover such data are important with respect to our understanding of mangrove ecosystem processes and the role of micro-organisms in maintaining these processes (Staley and Gosink, 1999).

3.1. Algae

Algae in the mangroves occur as epiphytes on the stems and roots of mangrove trees or growing on other substrata within the mangrove ecosystem. They are the main food source for a variety of fishes and invertebrates, such as crabs. Some algae are unique to certain mangrove habitats and an understanding of their diversity and biomass indicate the health of the mangroves. Algal mats (consisting largely of *Cyanophyta* and *Chlorophyta*) associated with mangrove habitats play an important role in building and trapping sediments or in carbonate precipitation (Hoffmann, 1999). Furthermore, many mangroves associated *Cyanophytes* (e.g. *Scytonema*) are important nitrogen fixers. Roots (pneumatophores) of the mangrove trees provide a favorable habitat for 50% of the total algal species, the hard substrates for 30% and the soft mud for 20% of the species (Kathiresan and Quasim, 2005). Certain algae are frequently associated with mangroves which are considered as characteristic of the ecology. The algal species belonging to the genera *Bostrychia, Cologlossa* and *Catenella* are commonly present on the roots and trunks of the mangroves. The species of the genera *Rhizoclonium, Enteromorpha* and *Cladophora* normally exist in the sediment along with the cyanobacteria (*Lyngbya* and *Anacystis*), benthic diatoms and the sulphate-reducing bacteria (Kathiresan and Quasim, 2005). This microbial community forms a 'biofilim" which activates the attachment of algae to the mangrove trees. The factors governing the occurrence of these algae are mainly those of a microclimate prevailing in the swamps. Most of these algae are small filament forms and are fairly resistant to desiccation as well as high salinity.

Algae are wide spread in the mangrove habitats all over the world. Cordeiro-Marino *et al.* (1992), recorded 150 algal taxa from the new world mangroves with highest diversity in the red algal groups with 78 species and lowest in the brown algae with less than 15 species. Maximum algal diversity with 109 species and high degree of endemism (about 70%) of the red algae was reported along the Caribbean coast (Kathiresan and Quasim, 2005). In Indian context, about 558 species of algae of 7 families were known to occur (Kathiresan and Quasim, 2005). The east coast is represented by 264 species, where as the west coast has 326 species (Kathiresan and Quasim, 2005). Only 71 species have been recorded so far from the Andaman and Nicobar Islands (Kathiresan and Quasim, 2005). From the Sundarbans mangroves of West Bengal, 150 different species of algal flora were identified belongs to different groups viz., *Chlorophyta* (39 species), *Chrysophyta* (44 species), *Phaeophyta* (2 species), *Rhodophyta* (15 species) and *Cyanophyta* (50 species) (Sen and Naskar, 2003). Algal species such as *Gloeocapsa* sp., *Chlorella* sp., *Ulva* sp., *Anabaena* sp., *Spirogyra* sp., *Oscillatoria* sp., *Phormidium* sp. were identified from mangroves of Bhitarkanika, Odisha. (Thatoi *et al.*, 2012).

3.1.1. Algal flora from mangrove environments of Odisha coast

Algal species have been collected from the mangrove soils, mangrove water and other hard substrates like tree bark, pneumatophores and submerged woods. Pure cultures of algae were made in the laboratory for their identification. The camera Lucida sketches and the photomicrographs of the algal species isolated from Bhitarkanika and Mahanadi delta mangrove environment were used for their identification purpose. The details of algal species identified from these mangroves were presented in the Table 10 and Fig. 19. So far, eleven algal genera viz., *Spirogyra* sp., *Anabaena* sp., *Gloeocapsa* sp., *Oscillatoria* sp., *Chlorella* sp., *Ulva* sp., *Chlorococcus* sp, *Chlamydomonas* sp., *Phormidium* sp., *Microcystis* sp. and *Lyngbya* sp, (*L. cylanica* and *L. semiplena*) have been reported from Bhitarkanika. Similarly in total, 6 genera of algae viz. *Spirogyra* sp., *Anabaena* sp., *Gloeocapsa* sp., *Oscillatoria* sp., *Chlorella* sp. and *Ulva* sp. were identified from mangroves of Mahanadi delta (Table 10). The identifying characteristics of individual algal species are given below.

Table 10: Algal species identified from mangroves of Odisha coast.

Sl. No.	Algal species	Place of occurrence	Reference
1	*Spirogyra* sp., *Anabaena* sp., *Gloeocapsa* sp., *Oscillatoria* sp., *Chlorella* sp., *Ulva* sp., *Chlorococcus* sp, *Chlamydomonas* sp., *Phormidium* sp., *Microcystis* sp. *Lyngbya* sp.	Bhitarkanika mangrove	Thatoi *et al.*, 2012
2	*Spirogyra* sp., *Anabaena* sp., *Gloeocapsa* sp., *Oscillatoria* sp., *Chlorella* sp. and *Ulva* sp.	Mahanadi mangrove	Thatoi *et al.*, 2010

1. *Spirogyra* sp.

The Spirogyra sp. is free floating, the large green silky and hair like cylindrical filaments are about 1/10 mm across and several centimeters long. Each cell contains 1-16 ribbon shaped chloroplasts arranged spirally in an anti-clockwise manner. A number of pyrenoids are present in the chloroplast. Each cell is longer than breadth and cylindrical with a two larged wall and uni-nucliated. The nucleus is suspended in the middle of the central vacule by cytoplasmic strands.

2. *Chlorella* sp.

It is a small, spherical, green unicellular alga found in fresh water as well as brackish water bodies. The alga is a unicellular which at most may grow to 10μm in diameter but usually it is much smaller. The small cells are non-motile, round and oval, usually found solitary, sometimes in groups. Chloroplast which is cup shaped or bell-shaped having parietal in position. Nucleus is present in the centre. Pyrenoids are usually absent. Each cell is field with dense cytoplasm.

3. *Oscillatoria* sp.

Trichome is single or forming a flat or spongy free-swimming thallus. Sheath is absent; rarely with a more or less very delicate sheath, motile, mostly by a creeping movement causing rotation on longitudinal axis. End of trichome is distinctly marked, pointed, but like a sickle or coiled; more or less like a screw. Hormogones are formed by the division of the trichome. Heterocyst is absent.

4. *Gloeocapsa* sp.

Cells are single with fewer colonies with distinct vesicular individual sheath. Sheath was uni-lamellated and colourless. Cells were spherical, 2-8 colonies, seldom many, with a number of concentric spherical envelopes. Colonies were single, individual sheaths were lamellated and cell division varied regularly in three directions.

5. *Ulva* sp.

The cells are simple with thick broad lobes. Blades are not naturally perforated, cells in transverse sections nearly square or longer than broad, small sized thallus, lobbed, blades were taper to form hold fast, cells nearly similar, irregularly arranged.

6. *Anabaena* sp.

Trichomes are uniformly broad throughout or apices may be somewhat attenuated. Sheath is absent. If present then form a soft mucilaginous thallus, heterocysts generally intercalary, spores single or in long series, formed near the heterocysts or in between the heterocysts. Akinetes were ellipsoidal with rounded ends.

7. *Chlamydomonas* sp.

It is motile single celled green alga about 10 μm diameter that swims with two flagella, It has a large cup-shaped chloroplast, a large pyrenoid, and an "eyespot" that senses light. Cells of this species are haploid having more or less oval cells having a cellulose membrane (theca), a stigma (eyespot); and a usually cup-shaped, pigment-containing chloroplast.

8. *Chlorococcus* sp.

The order includes unicellular, coenocytes and colonial, non-motile green algae. The colonial forms consist of a definite number of non-motile cells arranged in a specific manner. Motility is confined to the gametes and zoospores only. Vegetative division of the cell is absent. Division takes place only at the time of reproduction.

9. *Phormidium* sp.

The filaments are long, cylindrical, and may be curved or spiral. Thin, firm, colorless sheaths adhere closely to the trichomes. The filaments move by gliding, creeping, rotating, or oscillating both inside and outside of the sheaths. The cells are rectangular and have unconstricted or slightly constricted cross walls. The apical cells may have calyptra, and are more pointed, narrow, or rounded than the other cells. Usually, the thylakoids are arranged radially within the cell, creating a net-like or banded appearance. Gas vesicles are absent.

10. *Microcystis* sp.

Cells are spherical or elongated, many is in spherical; ellipsoidal or irregularly overlapping or net-like colony, free swimming, often with attached daughter colonies, cells are in homogenous colourless, often diffluent and mucilage. Individual envelops are absent. Cells mostly very densely arranged, generally transverse in elongate cells; gas-vacuoles often present.

11. *Lyngbya cylanica*

Thallus is olive - green, violet or red, filaments, 10-14 μm broad, straight sheath thin, colourless, trichome, blue - green or violet, 8- 12 μm broad, cross - walls not granulated, cell quadrate to 1/2 or 1/3 as long as broad, end cell round, without calyptras.

12. *Lyngbya semiplena*

The alga is mostly yellowish, green, some time when dry becomes dark. Filaments at the base become decumbent. When old the filaments were lamellated up to 3 μm thick, not coloured. Sometimes growing erect by their filament and sometimes entangled and curved.

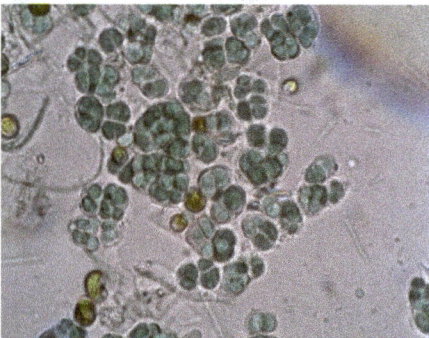

(a) *Oscillatoria* sp.   (b) *Gloeocapsa* sp.

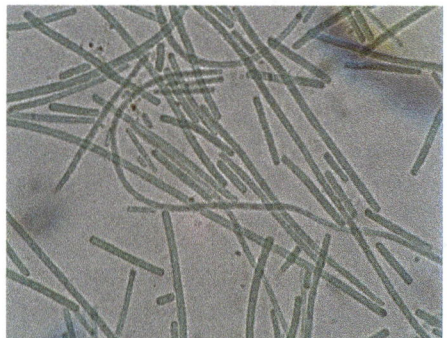

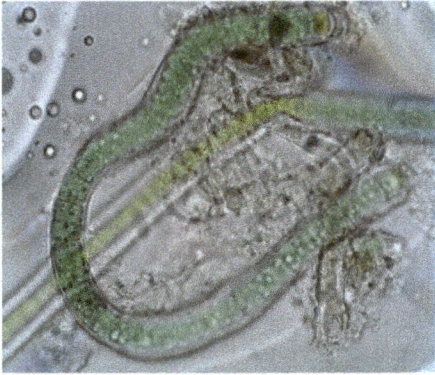

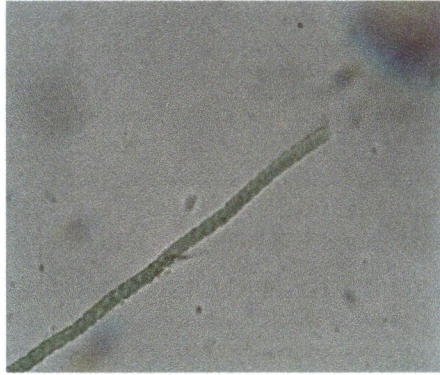

(c) *Anabaena* sp.	(d) *Phormidium* sp.

Figure 19: Phase contrast photograph of some algal species identified form Bhitarkanika mangrove ecosystem of Odisha. (a) *Oscillatoria* sp., **(b)** *Gloeocapsa* sp. **(c)** *Anabaena* sp. and **(d)** *Phormidium* sp.

3.2. Fungi

Mangrove forests are biodiversity 'hotspots' for marine fungi (Shearer *et al.*, 2007). Fungi are found associated with mangrove plants. The mangrove trunks and aerating roots are permanently or intermittently submerged, the upper parts of the roots and trunks rarely or never wetted by the salt water. Thus, terrestrial fungi and lichens occupy the upper part of the trees and marine species occupy the lower part and at the interface there is an overlap between marine and terrestrial fungi (Sarma and Hyde, 2001). Since they were first reported from mangrove roots in Australia by Cribb (1995), there has been considerable increase in information on mangrove associated fungi. The latest estimate of marine fungi in world is 1500 species, which excludes the lichens and many fungi those are newly isolated or inadequately described (Hyde *et al.*, 1998). Among the early studies, Hyde listed only 120 species of fungi from 29 mangrove forests around the world (Hyde, 1990). These included 87 *Ascomycetes*, 31 *Deuteromycetes* and 2 *Basidiomycetes*. There are some 169 fungal species from Malaysia (Alias *et al.*, 1995), 44 fungi associated with standing senescent *Acanthus ilicifolius* from Mai Po mangrove, Hong Kong (Sadaba *et al.*, 1995), 76 species from Pearl River Estuary, China (Vrijmoed *et al.*, 1991), 91 fungi from Egyptian Red Sea and 112 species from Bahamas islands (Abdel-Wahab, 2005) were reported.

Endophytic fungi were found in large number in mangrove environment. More than 200 species of endophytic fungi were isolated and identified from mangroves which are mainly *Alternaria, Aspergillus, Cladosporium, Clolletotrichum, Fusarium, Paecilamyces, Penicillium, Pestalotiopsis, Phoma, Phomopsis, Phyllosticta* and *Trichodema* (Liu *et al.*, 2007). Researchers from China, surveyed and reported the occurrence of several arbuscular mycorrhizal fungi (AMF) on root system of mangroves plants in QinZhou Bay, Guangxi, China (Wang *et al.*, 2003). In India, twenty-five endophytic fungi comprised of three ascomycetes, 20 mitosporic fungi and two sterile fungi were recovered from two halophytes (*Acanthus ilicifolius* and *Acrostichum aureum*) of a west coast mangrove habitat. Endophytic fungi were also isolated from leaves of *Rhizophora apiculata* and *Rhizophora mucronata*, two typical mangrove plants grown in the Pichavaram mangrove of Tamil Nadu, Southern India (Suryanarayanan *et al.*, 1998). A marine phosphate-solubilizing fungus, *Aspergillus niger* together with several phosphate-solubilizing bacterial strains, from the rhizosphere of black mangrove *Avicennia germinans* was reported by Vazquez *et al.* (2000). Similarly, Arbuscular mycorrhizal (AM) colonising fungi from river estuary of Ganges have been reported by Sengupta and Chaudhuri (2002).

In Indian context Ravikumar and Vittal, (1996) has reported occurring of 48 fungal species from decomposing *Rhizophora* debris in Pichavaram, South India. Seven species of fungi that exist on mangrove leaf surface of Sundarbans of West Bengal have been reported by Pal and Purkayastha, (1992). From Mangalvan mangrove ecosystem, 31 fungal isolates were recorded form soil and 27 species from decaying mangroves and 7 species from floating plants. Among these the dominant fungal species were *Aspergillus* followed by *Penicillium, Fusarium* and *Trichoderma*. Sarma &Vittal (2001) reported 73 species of fungi from Krishna estuaries of India. Similarly, 31 fungal species has been studied from sediment and 27 species from decaying leaves, stems, roots and pneumatophores of an estuarine mangrove ecosystem of Cochin (Prabhakaran and Gupta, 1990). Raghukumar *et al.* (1995) has studied the colonization *Thraustochytrids* on leaf litter of *Rhizophora apicula* at Caorao mangrove, Goa. Higher groups of fungi have been reported from the mangrove woods in Maharashtra coast with 41 species of *Ascomycetes*, 2 *Basidiomycetes*, and 12 *Deuteromycetes* with predominance of *Massarina velatospora* (Borse, 1988).

From mangroves of Bhitarkanika, Gupta *et al.* (2009) has reported the population status of fungi associated with the phyllosphere of different mangrove plants viz., *Avicennia, Aegiceras, Bruguiera, Ceriops, Excoecaria, Heritiera, Kandelia, Rhizophora* and *Sonneratia*. There have been 33 fungi from Godavari and 67 fungi from Krishna estuary, India were reported by Venkateswara *et al.* (2001). Relatively few fungi have been reported as pathogens of mangrove plants, as compared to the number of saprophytic fungi identified on decaying mangrove wood and leaves (Hyde *et al.* 1998). The intertidal fungus *Cytospora rhizophorae* is thought to be parasitic on *Rhizophora* spp. prop root (Kohlmeyer and Kohlmeyer, 1979). *Phomopsis mangrovei*, which is probably pathogenic, was described from dying prop roots of *Rhizophora apiculata* in Thailand (Hyde, 1996). An intertidal *Phytophthora* species was described to cause terminal dieback of *Avicennia marina* (Pegg *et al.*, 1980) and a *Phytophthora* species was also found to be pathogenic on *Avicennia marina* var. *resinifera* in New Zealand (Maxwell, 1968). *Halophytophthora* species were also thought to be responsible for diseased mangrove forests over vast areas in Sydney (Garrettson-Cornell and Simpson, 1984).

3.2.1. Fungi from mangroves of Odisha coast

Fungal species from mangrove forests of Odisha coast have been studied by Thatoi *et al.*, (2012) and Behera *et al.*, (2012). Fungal species have been isolated and identified from soil samples of four different locations of Bhitarkanika and Mahanadi mangrove forest. In total, 40 fungal species were identified from soil samples from these two mangrove areas (Table 11 and Fig. 20). Apart from their identification, distribution of individual fungal species in different soils has also been studied. Among the different fungal species identified, the population of *Fusarium solani* was highest followed by *Aspergillus oryzae* and *Fusarium oxysporum* in Bhitarkanika mangrove forest (Table 11). Besides the above, some arbuscular mycorrhizal fungi such as *Glomus, Acaulospora, Gigaspora, Scutellospora* and *Enterophospora* were recorded from three salinity zones of Bhitarkanika mangrove ecosystem (Gupta *et al.*, 2016).

Table 11: Ennumeration and identification of fungal species from mangroves of Bhitarkanika and Mahanadi delta, Odisha

Sl. No.	Name of the fungi	Bhitarkanika	Mahanadi Delta
		cfu/g.soil x 10⁻⁴	cfu/g.soil x 10⁻⁴
1	Acremonium roseum	ND	15-24
2	Alternaria alternata	31-67	23
3	Aspergillus albus	04	ND
4	Aspergillus flavus	35	20
5	Aspergillus niser	58-66	23
6	Aspergillus oryzae	78-121	22
7	Cladosporium oxysporum	24	10
8	Epicoccum purpura	ND	10
9	Fusarium oxysporum	43-69	13
10	Fusarium pallidoroseum	ND	2
11	Fusarium solani	35-168	39-50
12	Graphium spp.	ND	08-24
13	Geotrichum candium	ND	04
14	Verticillium spp.	ND	16
15	Paecilomyces variotii	20	ND
16	Penicillium chrysogenum	ND	20
17	Penicillium citrinum	6-32	19
18	Penicilliumred sclerotium	ND	36
19	Phoma spp	ND	22-27
20	Streptomyces albus	ND	21-25
21	Trichoderma Virnse	ND	10-55
22	Trichoderma viride	25-47	150
23	Acremonium byssoides	51	ND
24	Choenophora ucrbitarum	5	ND
25	Curvularia lunata	35	ND
26	Drechslera hawaiensis	45	ND
27	Fusarium chlamydosporum	10	ND

Sl. No.	Name of the fungi	Bhitarkanika	Mahanadi Delta
28	Fusarium moniliforme	36	ND
29	Fusarium redolense	45	ND
30	Neocosmospora asinfectum	0-4	0-7
31	Neosortaria fischeri	0-7	ND
32	Paecilomyces lilacinous	0-3	ND
33	Paecilomyces varioti	0-2	ND
34	Phoma glomerta	2-5	2-3
35	Penicillium digitatum	0-5	ND
36	Penicillium oxalicum	0-9	5-7
37	Rhizopus stolonifer	10-14	7-9
38	Streptomyces species	10-23	7-11
39	Trichoderma harzianum	24-65	12-32
40	Byssochlarnus niveus	1-3	ND

Note: ND= not detected

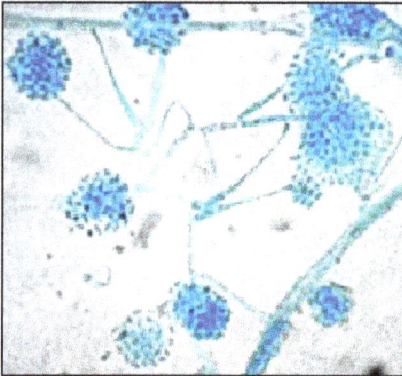

1. *Acremonium byssoides*

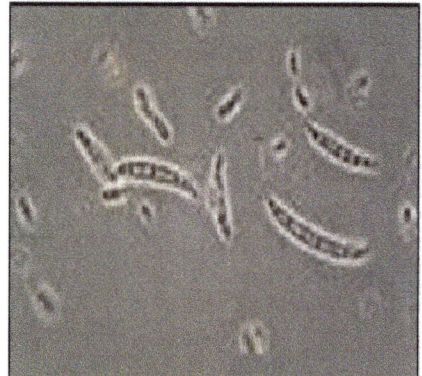

2. *Alternaria alternata*

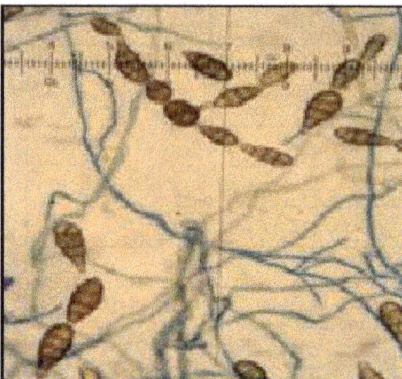

3. *Aspergillus glavus*

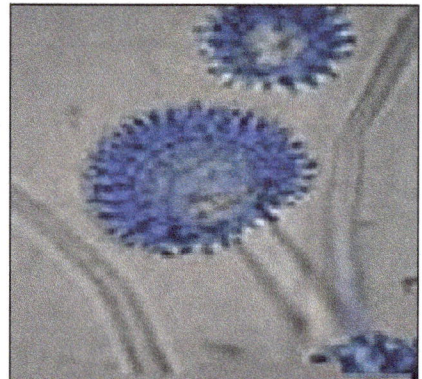

4. *Aspergillus niger*

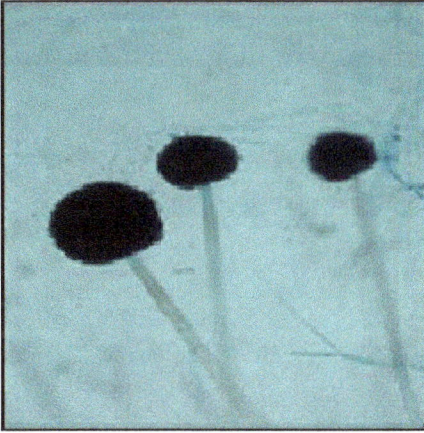

5. *Aspergillus oryzae*

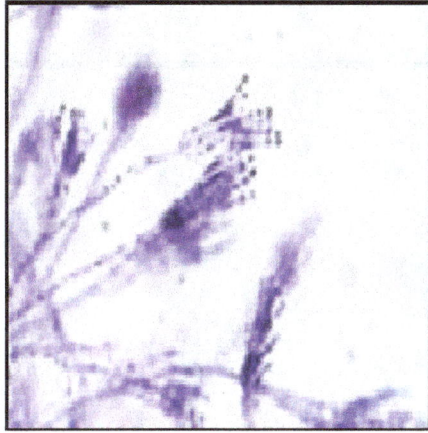

6. *Choenophora cucrbitarum*

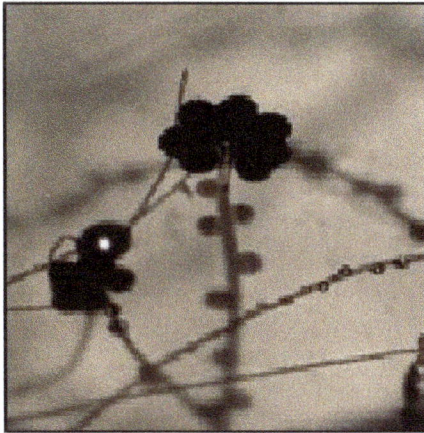

7. *Cladosporium oxysporum*

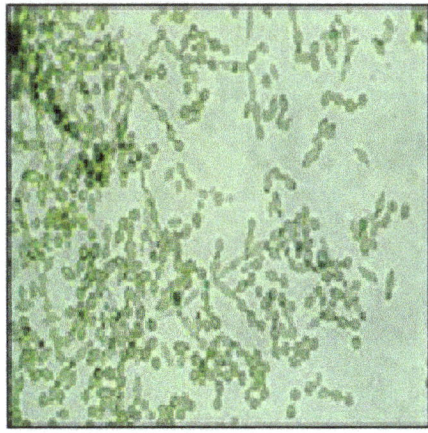

8. *Curvularia lunata*

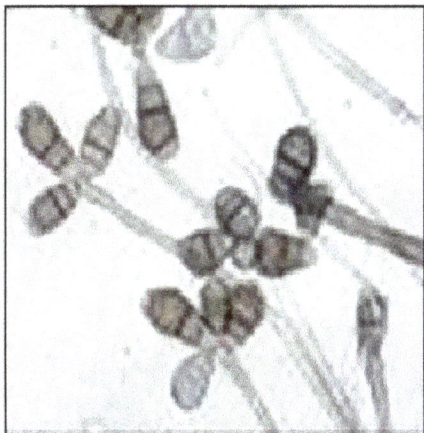

9. *Drechslera hawaiensis*

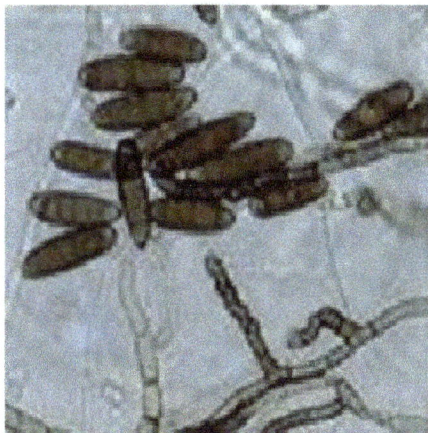

10. *Fusarium oxysporum*

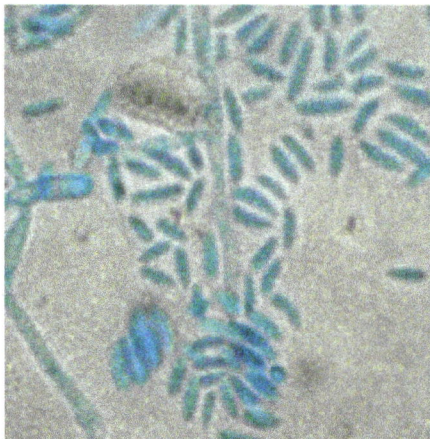

11. *Fusarium moniliforme*

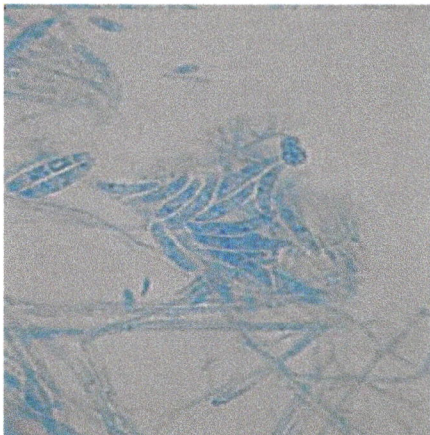

12. *Fusarium solani*

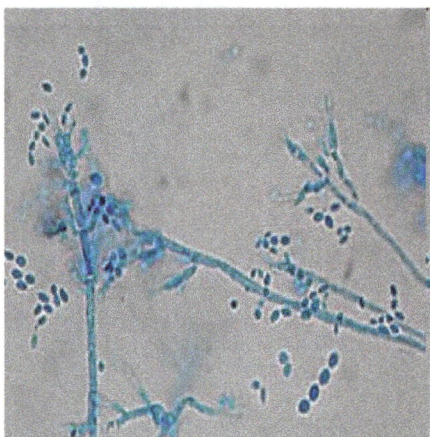

13. *Paecilomyces lilacinous*

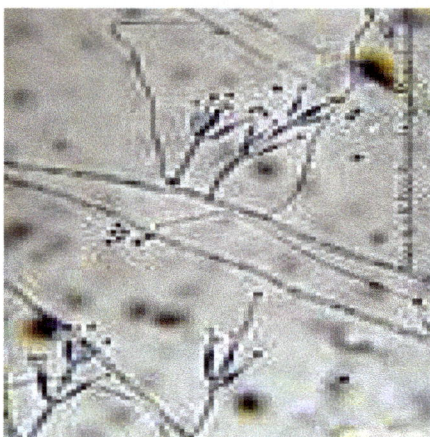

14. *Paecilomyces varioti*

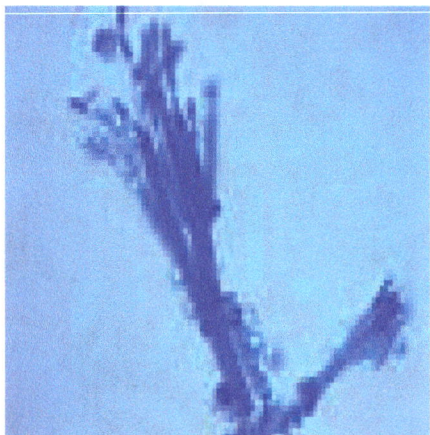

15. *Penicillium digitatum*

16. *Trichoderma harzianum*

17. *Penicillium citrinum*

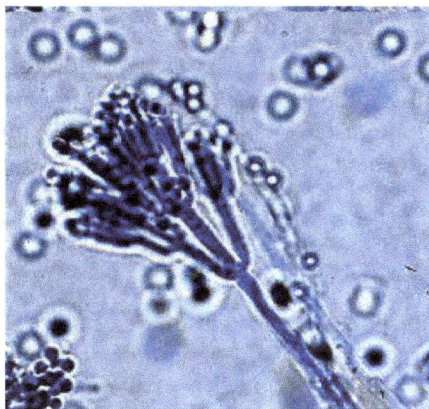
18. *Penicillium oxalicum*

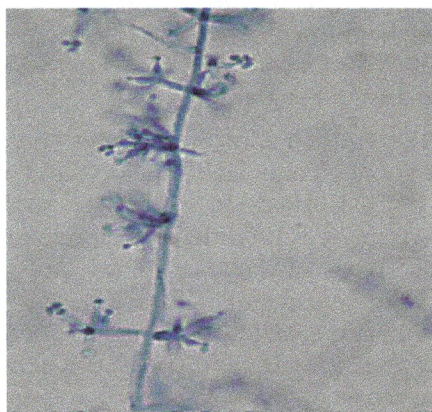
19. *Rhizopus stolonifer*

20. *Trichoderma viride*

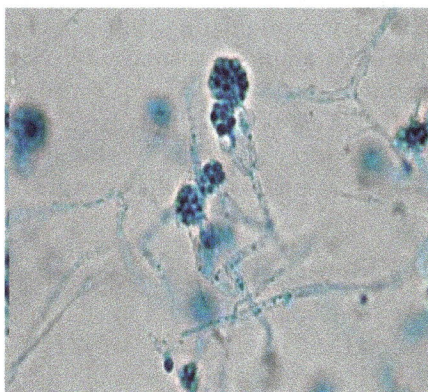
21. *Trichoderma virense*

Figure 20: Fungal species identified from mangrove soils of Bhitarkanika and Mahanadi delta, Odisha

3.3. Actinomycetes

Actinomycetes have been looked upon as potential sources of bioactive compounds and are the richest sources of secondary metabolites. The mangrove ecosystem is a largely unexplored source for actinomycetes with the potential to produce biologically active secondary metabolites (Hong *et al.*, 2009). Several reporters from different geographical locations of the world have described the occurrence of actinomycetes in different mangrove habitats of the world. Eccleston *et al.* (2008) reported the occurrence of actinomycetes belongs to genus *Micromonosporae* from Sunshine Coast in Australia. Rifamycin producing *Micromonospora* from mangrove of South China Sea has been reported by Huang *et al.* (2008) and Xie *et al.* (2006). Several genera of actinomycetes such as *Actinomadura, Microbispora, Nonomuraea, Actinoplanes, Micromonospora, Verrucosispora, Arthrobacte, Isoptericola, Micrococcus, Microbacterium, Nocardia, Rhodococcus* and *Streptomyces* were reported from mangrove soils and plants in china (Hong *et al.*, 2009). Similarly, genera like *Brevibacterium, Dermabacter, Kocuria, Kytococcus, Microbacterium, Nesterenkonia* and *Rothia* were reported from mangrove sediment of Brazil (Dias *et al.*, 2009). Ara *et al.* (2007) have reported novel actinomycetes (*Nonomuraea maheshkhaliensis*) from a mangrove rhizosphere mud in the southern area of Bangladesh.

From Indian context, Sivakumar (2001) reported 23 actinomycetes species from Pichavaram mangrove and most of the species identified belongs to the genus *Streptomyces*. Laksmanaperumalsamy *et al.* (1978) isolated 518 *Streptomyces* strains from mangrove environment of Porto Novo. As many as 107 different actinomycetes were isolated from marine sediments of Konkan coast of Maharashtra by Gulve and Deshmukh (2011) and 17 actinomycetes isolates were identified from Karangkadu mangrove forest Tamil Nadu, India by Ravikumar *et al.* (2010). Sahu *et al.* (2005) reported several *Streptomycetes* viz., *S. alboniger, S. violaceus, S moderatus* and *S. aureofasciculus* from Vellar estuary south east coast of India. Similarly Ravikumar *et al.* (2011) have reported the biodiversity of actinomycetes from the sediments of Manakkudi mangrove ecosystem of southwest coast of India. Sivakumar *et al.* (2005c) reported the occurrence of *S. albidoflaus* from the Pichavaram mangrove which has antitumor properties. Phosphate solubilizing actinomycetes such as *Streptomyces galbus* in the Vellar Estuary of Parangipettai estuarine environment, south east coast of India has been reported by Sahu *et al.* (2007d). Distribution of actinomycetes in the Sundarbans mangrove of West Bengal, India has been reported by Mitra *et al.* (2008). Relatively large distribution of actinomycetes species all over the world's mangrove ecosystem appears to be reasoning that mangrove forests are the treasure house for actinomycetes.

3.3.1. Actinomycetes identified from mangroves of Odisha coast

Actinomycetes study from mangrove of Odisha coast is limited which are confined to Bhitarkanika and Devi river estuary only (Table 12). Gupta *et al.* (2009) has reported a number of *Streptomyces* such as *S. albidoflavus, S. atroolivaceous, S. auranticus, S. canus, S. chromofuscus, S. exfoliates, S. griseoluteus, S. helstedii, S. lavenduale, S. longisporoflavus, S. luridus, S. lydicus, S. nogalator, S. pactum, S. prasinosporus, S. purpureus, S. tubercidus, S. versoviensis, S. viridochromogenes* and *S. xanthochromogenes* from different plant species of Bhitarkanika mangrove forest of Odisha. Five phosphate solubilising *Streptomyces* sp. from mangrove ecosystem of Bhitarkanika, Odisha, were also reported by Mohanta *et al.* (2014). Seven actinomycetes sp. such as *Streptomyces, Sacharopolyspora, Nocardiopsis, Micromonospora, Actinomadura,* and *Actinopolyspora* were reported from Bhitarkanika mangrove by Rajkumar *et al.,* (2012). Actinomycetes sp. such as *S. almquisti, S. vastus, S. alni, S. luteogriseus, A. longiporus, A. malachitorectus, S. neyagawaensis, A. aureocirculatus, A. janthinus, S. spheroides, S. albulus, S. antibioticus, S. mirabilis, S. umbrosus, S. thermovulgaris* were also reported from Bhitarkanika mangrove by Kishore (2011). Similarly *Streptomyces sampsonii, Streptomyces flavogreseus* were reported from Devi estuary of Odisha by Sahu *et al.,* (2013).

Table 12: Actinomycetes identified from mangroves of Odisha coast

Mangrove	Actinomycetes identified	References
Bhitarkanika		
	Streptomyces sp., Sacharopolyspora sp., Nocardiopsis sp., Micromonospora sp., Actinomadura sp., Actinomycetes sp., Actinopolyspora sp.	Rajkumar *et al.,* (2012)
	S. almquisti, S. vastus, S. alni, S. luteogriseus, A. longiporus, S. neyagawaensis, A. aureocirculatus, A. janthinus, S. spheroides, S. albulus, S. antibioticus, S. mirabilis, S. umbrosus, S. thermovulgaris	Kishore, (2011)
	Streptomyces sp.,	Mohanta *et al.,* (2014)
	S. albidoflavus, S .atroolivaceus, S. aurantiacus, S. canus, S. chromofuscus, S. exfoliates, S. griseoluteus, S. helstedii, S. lavenduale, S. longisporoflavus, S. luridus, S. lydicus, S. nogalator, S. pactum, S. prasinosporus, S. purpureuss, S. tubercidicus, S. varsoviensis S. viridochromogenes, S. xanthochromogenis	Gupta *et al.,* (2009)
Devi estuary	*Streptomyces sampsonii, Streptomyces flavogreseus*	Sahu *et al.,* (2013)

3.4. Bacteria

Next to trees bacterial flora dominates the biomass and productivity of mangrove forests (Kathiresan and Quasim, 2005). Among the microbes, the bacterial populations in mangroves are many fold greater than those of the fungi (Kathiresan and Quasim, 2005). Microbial generated detritus in mangrove ecosystems acts as the major substrate for bacterial growth in mangroves (Bano *et al.,* 1997). The bacteria may act as primary decomposers, which utilize dissolved organic substances at low concentration and assimilate dissolved inorganic substances like nitrate and phosphate. Different groups of bacteria in mangrove ecosystem perform different ecological role such as nitrogen fixation, phosphate solubilization, cellulose degradation and sulfur oxidation. The bacteria exist as symbionts with plants and animals, saprophytes on dead organic matter and as parasites on living organisms (Kathiresan and Quasim, 2005). The bacteria performed varied activities in the mangrove ecosystems like photosynthesis, nitrogen fixation, methanogenesis, magnetic behaviour, human pathogens, production of antibiotics and enzymes. Some major groups of bacteria in mangrove ecosystem are discussed below.

3.4.1. Photosynthetic anoxygenic bacteria

Mainly two types of photosynthetic bacteria are seen in mangrove ecosystem such as purple sulphur bacteria (family Chromatiaceae,) and purple non-sulphur bacteria (family Rhodospirillaceae, strain belonging to *Rhodopseudomonas* sp.). These bacteria are capable of using light to grow, fix nitrogen and release hydrogen gas in this environment. Purple sulphur bacteria (PSB) range in color from pink to purple and contain bacteriochlorophyll a as their major pigment. These phototrophic anaerobes require sulphide, which they oxidize to sulphate for growth. Carbon dioxide is the usual source of cell carbon, but they also utilize various organic acids as carbon sources and usually widely distributed in sulphide rich environment such as mangroves (Vethanayagam, 1991). Purple non sulphur bacteria (PNB) range in color from brown to red and also contain bacteriochlorophyll a as their major pigment. They have the ability to utilize remarkably wide spectrum of reducing carbon compounds, like malate or succinate as an electron donor as well as carbon sources for growth. Sulphur rich mangrove ecosystem, which is mainly anaerobic soil environment, provides favorable conditions for the proliferation of these bacteria. The predominant bacteria belongs to this group in the mangrove ecosystem of Cochin (India) were identified as the member of the genera *Chloronema, Chromatium, Beggiatoa, Thiopedia* and *Leucothiobacteri* (Vethanayagam and Krishnamurthy, 1995). In mangrove on the coast of Red Sea in Egypt, 225 isolates of purple non-sulfur bacteria, belonging to ten species of four different genera, were identified. Nine of the ten species inhabited the rhizosphere and root surface of the trees. The most common bacteria *Rhodobacter* and *Rhodopseudomonas* were detected in 73% and 80% of the sample respectively (Shoreit *et al.*, 1994). Some of the anoxygenic photosynthetic bacteria were also diazotrophic. Although there is yet no published evidence, one can hypothesized that photosynthetic anoxygenic bacteria, the predominant photosynthetic organisms in anaerobic environments, may contribute to the productivity of the ecosystems (Sahoo and Dhal, 2009).

3.4.2. N_2 fixing bacteria

Nitrogen fixation is a process of conversion of gaseous forms of Nitrogen (N_2) into combined forms i.e. ammonia or organic nitrogen by some bacteria and cyanobacteria. Free living as well as symbiotic microbes known as diazotrophs which fix N_2 into proteins. Nitrogen fixing (diazotrophic) microorganisms can colonize both in terrestrial and marine environments. In mangrove ecosystems, high rates of nitrogen fixation has been associated with dead and decomposing leaves (Mann and Steinke, 1992), pneumatophores (Hicks and Silvester 1985; Toledo *et al.*, 1995a) and the rhizosphere soil (Holguin *et al.*, 1992). N_2 fixation in mangrove sediments is likely to be limited by insufficient energy sources. The low rates of N_2 fixation by heterotrophic bacteria detected in marine water are probably due to lack of energy sources. Nitrogen fixation by heterotrophic bacteria can be regulated by specific environmental factors such as oxygen, combined nitrogen and the availability of carbon source to support energy requirement. Energy for N_2 fixation can also be derived from leaves and roots decomposed by non diazotrophic microflora that colonize dead mangrove leaves (Zuberer and Silver, 1978). Nitrogen-fixing bacteria such as members of the genera *Azospirillum, Azotobacter, Rhizobium, Clostridium* and *Klebsiella* were isolated from the sediments, rhizosphere and root surfaces of various mangrove species. Nitrogen-fixing bacteria, *Azotobacter species* are repeatedly isolated from sediments of Pichavaram mangroves and they were more in the mangrove habitats than in the back waters and estuarine systems (Lakshmanaperumalsamy, 1987). Several strains of diazotrophic bacteria such as *Vibrio campbelli, Listonella anguillarum, Vibrio aestuarianus* and *Phyllobacterium* sp. were isolated from the rhizosphere of the mangroves in Mexico (Holguin *et al.*, 1992). N_2 fixing bacteria such as *Azotobacter* sp. which can be used as biofertilizers, was abundant in mangrove habitats of Pichavaram (Ravikumar, 1995). Two halotolerant N_2 fixing *Rhizobium* strains were isolated from root nodules of *Derris scandens* and *Sesbania* species growing in the mangrove swamps of Sundarbans (Sengupta and Chaudhuri, 1990). Nitrogen fixing cyanobacteria such as *Aphanocapsa* sp., *Nodularia* sp. and *Trichodesmium* sp. were isolated from Pichavaram mangroves (Ramachandran and Venugopalan, 1987). N_2 fixing bacteria are efficient at using a variety of mangrove substrates despite differences in carbon content and phenol concentrations (Pelegri and Twilley, 1998). However, their abundance may be dependent on physical conditions and mangrove community composition. Both symbiotic and asymbiotic N_2 fixing bacteria play a vital role on nitrogen enrichment of the mangrove ecosystems (Holguin *et al.*, 2001). One may conclude from the available information that N_2 fixation is a major bacterial activity in mangrove ecosystems, second only to carbon decomposition of detritus by the sulfate-reducing bacteria.

3.4.3. Phosphate solubilising bacteria

Phosphorous is one of the major plant nutrients, second to nitrogen (Vassileva *et al.*, 1998). So the phosphate solubilising microorganisms (PSMs) play an important role in supplementing phosphorus to the plants and allowing a sustainable use of phosphate fertilizers (Gyaneshwar *et al.*, 1998). Muddy mangrove soils have a strong capacity to absorb nitrates and phosphates carried by the tides (Hesse, 1962). Most of the inorganic phosphate present in the sediment is bound to calcium, iron and aluminium ions as insoluble phosphates (Alongi *et al.*, 1992). Fungi and inorganic phosphate solubilising bacteria present in the mangrove rhizosphere participate in releasing soluble phosphate into water (Vazquez *et al.*, 2000). Certain bacteria exhibit high phosphatase activity, capable of solubilising phosphate (Sundararaj *et al.*, 1974). In an arid mangrove ecosystem in Mexico, nine strains of phosphate-solubilizing bacteria such as *Bacillus amyloliquefaciens, B. atrophaeus, Paenibacillus macerans, Xanthobacter agilis, Vibrio proteolyticus, Enterobacter aerogenes, E. taylorae, E. asburiae* and *Kluyvera cryocrescens* were isolated from black mangroves (*Aviciena germinans*) roots. Further, three strains such as *B. licheniformis, Chryseomonas luteola* and *Pseudomonas stutzeri* were isolated from white mangrove (*Laguncularia racemosa*) roots (Vazquez *et al.*, 2000). A very little information is available about phosphate solubilising bacterial diversity and their activity in Indian mangroves. However, some studies related to phosphate solubilizing bacterial activity has been done by Kathiresan and Selvam (2006) from Vellar estuary at Parangipettai South eastern coast of India, Gupta *et al.* (2007) and Thatoi *et al.* (2012) from Bhitarkanika mangrove environment of Odisha, Ramnathan *et al.* (2008) from Sundarban mangroves of West Bengal and Kothamasi et al. (2006) from the Great Nicobar mangroves of India. The preliminary isolation and screening of phosphate solubilizing bacteria from mangrove soil of Bhitarkanika, Odisha cost was done by Gupta *et al.* (2007)

revealed the presence of 33 soil bacteria showing phosphate solubilizing capacity. Ramanathan *et al.* (2008) has quantified phosphorus solubilizing bacteria along with cellulose degrading and N_2 fixing bacteria from Sundarban mangroves of India. Kothamasi *et al.* (2006) has reported two strains of phosphate solubilizing *Pseudomonas aeruginosa* (designated GM01 and GM02) were found in the mangrove soils of Great Nicobar. Nine phosphate solubilizing bacteria has been isolated and phenotypically characterized from mangrove soil of Bhitarkanika by Mishra *et al.* (2010). Phosphate solubilizing bacteria like genera *Pseudomonas*, *Bacillus*, *Corynebacterium*, *Vibrio*, *Micrococcus* and *Alcaligens* were studied by Venkateswaran and Natarajan (1983) in mangrove biotopes in Porto Novo, Chennai water and sediment. Endophytic phosphate solubilising bacteria were isolated from leaf samples of mangrove plants of Pichavaram, Tamil Nadu by Gayatri *et al.* (2010).

3.4.4. Sulphur oxidizing bacteria

Mangrove sediments are mainly anaerobic with an overlying thin aerobic sediment layer. Degradation of organic matter in the aerobic zone occurs by various microorganisms and among various microorganisms, bacteria play major roles in the chemical and biological redox reactions in this ecosystem that create the biogeochemical cycle. Among the various biogeochemical cycles that takes place in this rich detritus based coastal sediment; the sulphur cycle is one of them. Sulphur oxidation improves soil fertility. It results in the formation of sulphate, which can be used by plants. Sulphur oxidizing bacteria play an important role in the detoxification of reduced sulphide in sediments.

Sulphate-reducing bacteria cycle hydrogen sulphide through the atmosphere for use by anaerobic photosynthetic bacteria and sulphur-oxidizing bacteria, while returning carbon dioxide to the atmosphere (Holmer *et al.*, 2001). Mangrove sediments are mainly anaerobic with an overlying thin aerobic sediment layer. Degradation of organic matter in the aerobic zone occurs principally through aerobic respiration whereas in the anaerobic layer decomposition occurs mainly through sulfate-reduction (Sherman *et al.*, 1998). Sulfur-oxidizing bacteria play an important role in the detoxification of sulfide in sediments. Symbiotic sulfur-oxidizers, e.g., those within members of the bivalve family Lucinacea, can be commonly found in muddy mangrove areas (Liang *et al.*, 2006). Sulfate reduction accounts for almost 100% of the total emission of CO_2 from the sediment (Kristensen *et al.*, 1991). Some of the sulphur oxidising bacteria such as Gammaproteobacteria e.g. *Chromatiales* and Deltaproteobacteria e.g. *Desulfobacterales* were reported from oil contaminated soil of Brazilian pristine mangrove sediment (Holguin *et al.* 2001; Santos *et al.*, 2011). Some of the free-living and symbiotic sulphur oxidising bacteria were reported from Futian mangrove swamp of China (Liang *et al.*, 2006). In Florida, sulfate-reducing bacteria were the most numerous bacterial group in the rhizosphere of *R. mangle* and *A. germinans* mangroves, reaching a population density of 10^6 cfu g^{-1} fresh weights (Zuberer and Silver, 1978). In Goa's mangrove (India), 10^3 cfu g^{-1} dry sediment of sulfate reducing bacteria, mostly spore-forming species, were associated with mangroves (Saxena *et al.*, 1988). Further, In Goa's mangroves, eight species of sulfate-reducing bacteria such as *Desulfovibrio desulfuricans*, *Desulfovibrio desulfuricans aestuarii*, *Desulfovibrio salexigens*, *Desulfovibrio sapovorans*, *Desulfotomaculum orientis*, *Desulfotomaculum acetoxidans*. *Desulfosarcina variabilis* and *Desulfococcus multivorans* were isolated and tentatively classified within four different genera (Lokabharati *et al.*, 1991). In mangrove sediments, availability of iron and phosphorus may also depend on the activity of sulfate reducing bacteria (Holguin *et al.*, 2001). It appears that sulfate-reducing bacteria, as the main decomposers of organic matter in anaerobic sediments which play a major role in the mineralization of organic sulfur and production of soluble iron and phosphorus which is used by the organisms in mangrove ecosystems.

3.4.5. Cellulose degrading bacteria

Cellulose is the primary product of photosynthesis in terrestrial environments and the most abundant renewable bio resource product in the biosphere (Zhang and Lynd, 2004a). Cellulose is a linear polysaccharide which is constructed from monomer of glucose bound together with ß1-4 glucosidal linkage. The transfer of carbon and energy from mangrove detritus to animal consumers appears to occur via grazing of easily digestible and highly nutrive microbial biomass resulting from bacterial and fungal transformation of the lignocellulosic detritus. Thus, most of the bacteria and fungi degrade the cellulosic material by producing enzyme called cellulase. Cellulose biodegradation by cellulases and cellulosomes, produced by numerous microorganisms, represent the major carbon flow from fixed carbon sinks to atmosphere CO_2 is very important in several agricultural and waste treatment processes (Haight, 2005). In anaerobic environment, which is rich in decaying plant material, decomposition of the cellulose is brought about by complex communities of interacting microorganisms (Odum and Heald, 1972). As the substrate i.e. cellulose is insoluble, bacterial and fungal degradation occurs exocellularly, to degrade cellulose into carbon and energy sources which is required for other microorganisms present in the mangrove environment. Several marine bacterial species such as *Rhodospirillum rubrum*, *Cellulomonas wmi*, *Clostridium stercorarium*, *Bacillus polymyxa*, *Pyrococcus furiosus*, *Acidothermus cellulolyticus* and *Saccharophagus degradans* were reported of degrading cellulose (Taylor *et al.*, 2006). Five promising cellulose producing bacteria such as *Bacillus cereus*, *Bacillus licheniformis*, and *Bacillus pumilus* and *Bacillus* sp. have been reported from philipines mangroves (Tabao and Monsalud, 2010). Sediment associated with dense Sunderban mangroves showed highest count of cellulose degrading bacteria in comparison to other bacterial diversity (Ramanathan *et al.*, 2008). In comparison to other bacterial diversity very less information is available on diversity of cellulose degrading bacteria from mangrove ecosystem, which may be due to the lack of suitable technologies for their isolation and identification. A detail study is required to assess the diversity of cellulose degrading bacteria from various mangrove ecosystems.

3.4.6. Methanogenic bacteria

One of the important characteristic of the mangrove sediments is the absence of oxygen at a few millimetres below the surface (Lyimo *et al.*, 2002). The lack of oxygen, coupled with the abundance of organic matter, creates an optimal environment for several groups of anaerobic organisms, such as sulphate-reducing bacteria (SRB) and methanogens (Dar *et al.*, 2008). Since the presence of these groups in most coastal sediments is selected by the redox potential (Dar *et al.*, 2008). These groups are expected to be found in discrete niches. However, these organisms are known to share similar niches in rich organic matter (OM) environments like mangroves

(Oremland *et al.*, 1982). Niche superimposition between SRBs and methanogens is restricted to certain substrates, such as hydrogen and acetate (Oremland *et al.*, 1982). Simple substrates (e.g., methanol, mono-di-trimethylamine) are important for methanogens (Lyimo *et al.*, 2002), but not for SRB, which are capable of degrading more complex substrates, such as long-chain and aromatic hydrocarbons (Muyzer and Stams, 2008). The presence of sulphate reducing bacteria limits the proliferation of these bacteria (Ramamurthy *et al*, 1990). There are several reports of occurring methanogenic bacteria in mangrove ecosystem such as, a strain of methanogenic bacterium, *Methanococcoides methylutens* (Mobanraju *et al.*, 1997) and four strain of unidentified thermotolerent methanogenic bacteria were isolated from the sediment of mangrove forest (Marty, 1985). A methanogenic bacterium, *Methanococcoides methyjuteus,* was isolated and characterised from the sediment of mangrove environment of Pichavaram, southeast of India (Mobanraju *et al.*, 1997). Lyimo *et al.* (2008) also reported methanogenic bacterium, *Methanococcoides methylutens* and *Methanosarcina semesiae* from sediment samples of Tanzanian mangrove and Taketani *et al.* (2010) reported the occurrence of methanogenic bacteria such as *Methanopyrus kandlery* and *Methanothermococcus thermolithotrophicus* from a pristine tropical mangrove soil of Brazil.

3.5. Ennumeration of different groups of bacteria from Bhitarkanika and Mahanadi mangroves of Odisha

Bacterial population such as heterotrophic, N_2 fixing, Gram (-) ve, nitrifying, sulphur oxidizing, Gram (+) ve, spore forming, denitrifying, phospahte solubilising, anaerobic, cellulose degrading bacteria were isolated and enumerated from different locations from Bhitarkanika and Mahanadi delta mangroves of Odisha using group specific media. Major groups of bacteria were identified with help of phenotypic and molecular techniques. Population of bacteria from Bhitarkaniaka and Mahanadi delta is presented in Table 13.

Table 13: Ennumeration of different groups of bacteria from mangroves of Odisha coast (Data is average of four seasons i.e. rainy, autumn, winter and summer of different sites)

Different groups of bacteria	Bacterial population (cfu/g.soil x 10^{-6})	
	Bhitarkanika mangrove	Mahanadi mangrove
Heterotrophic	138.00-413.00	46.00-167.00
Nitrogen fixing	91.00-281.00	28.00-113.00
Gram negative	44.00-175.00	8.00-134.00
Nitrifying	22.00-75.66	8.33 – 104.23
Sulphur oxidizing	7.04-41.33	5.60 – 28.00
Gram positive	44.00-118.00	9.33 – 110.00
Spore forming	11.00-46.05	5.30 – 60.23
Denitrifying	5.05-32.70	4.33 – 114.22
Anaerobic	4.00-24.33	2.10 – 12.22
Phospahte solubilising	2.22-15.66	1.25 – 48.60
Cellulose degrading	2.34-10.88	1.33 – 15.23
Actinomycetes	2.05-9.00	1.05-6.23

3.5.1. Ennumeration of bacterial population of Bhitarkanika mangrove forests

Enumerations of twelve different groups of bacteria from the soil samples of five different sites viz., Site-1 (Rangani), Site-2 (Mahisamunda), Site-3 (Habalaganda), Site-4 (Dangamal), Site-5 (Kalibhanjadian) of the Bhitarkanika mangrove ecosystem of Odisha during four seasons i.e. rainy, autumn, winter and summer are given in (Fig. 21a-21l). Population of all of the eleven different groups of bacteria i.e. heterotrophic, N_2 fixing, Gram (-)ve, nitrifying, S oxidizing, Gram (+)ve, spore forming, denitrifying, P solubilizing, anaerobic and cellulose degrading bacteria and actinomycetes did not change coherently either season wise or site wise (Fig. 21a-21l). Generally, the heterotrophic, P solubilizing (4-9 cfu/g soil) and S oxidizing (13-41.33 cfu/g soil) guilds were optimum in the rainy season, when most of the other guilds declined or maintained the pool size (Figs 21a-21l).

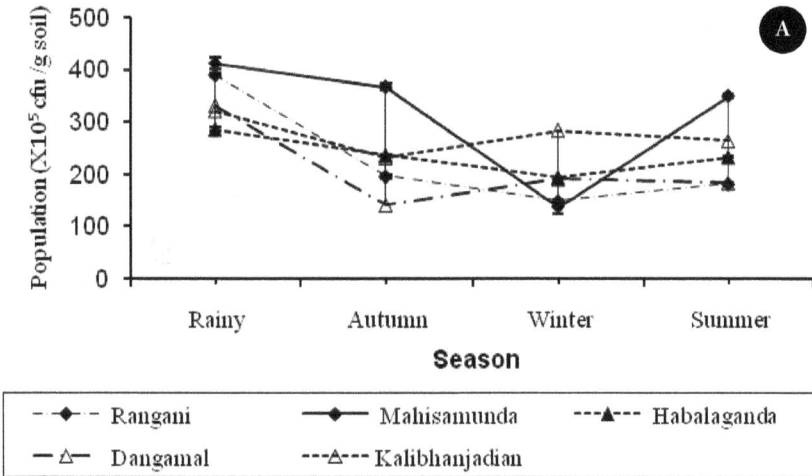

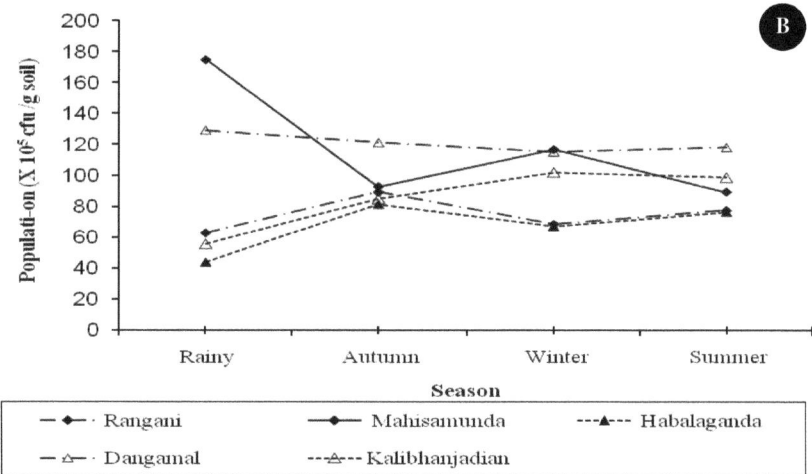

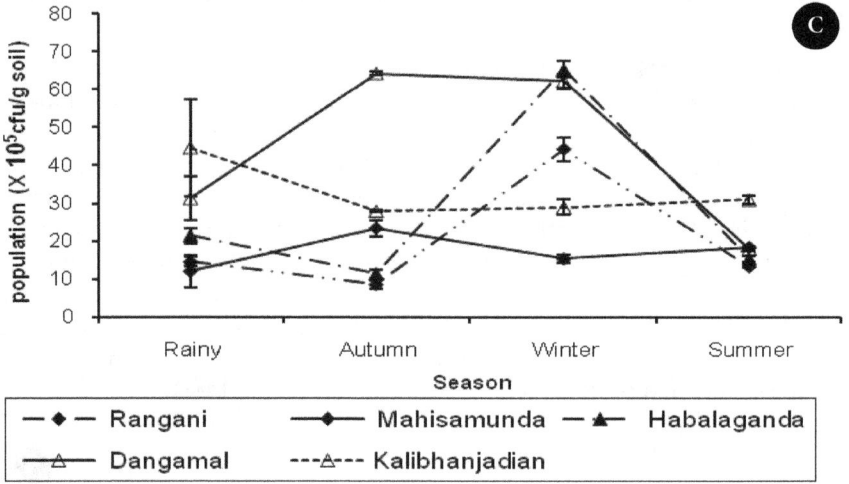

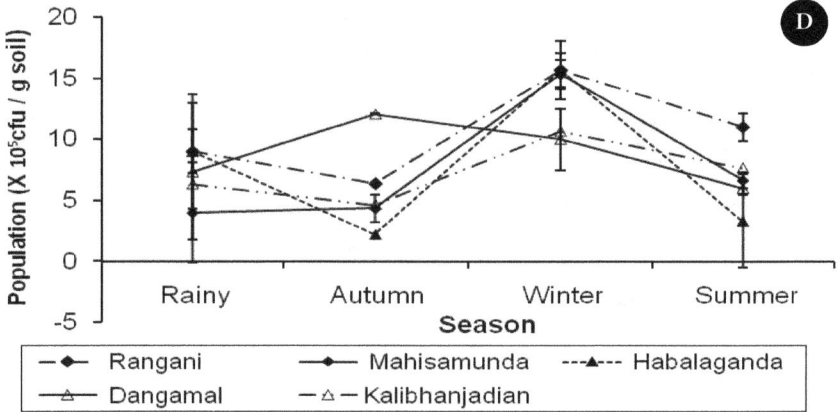

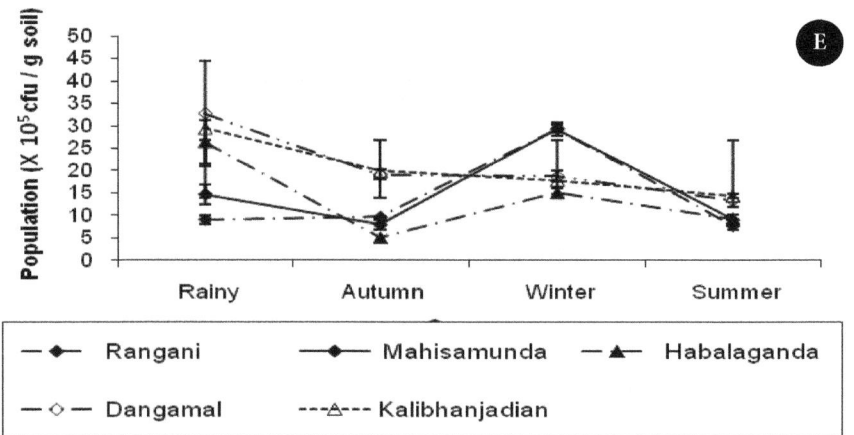

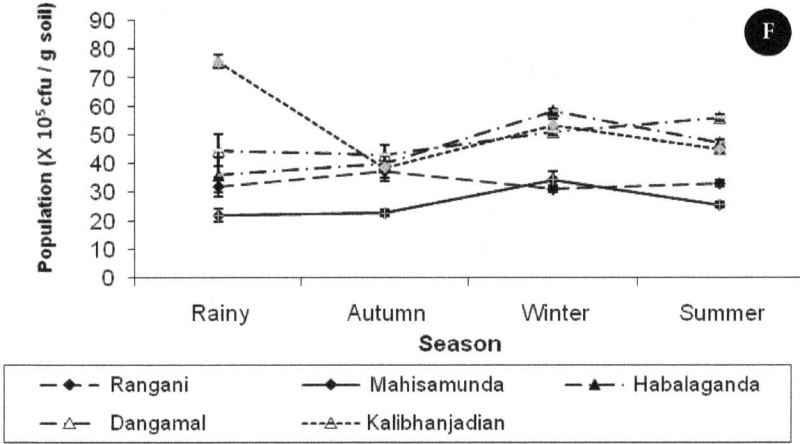

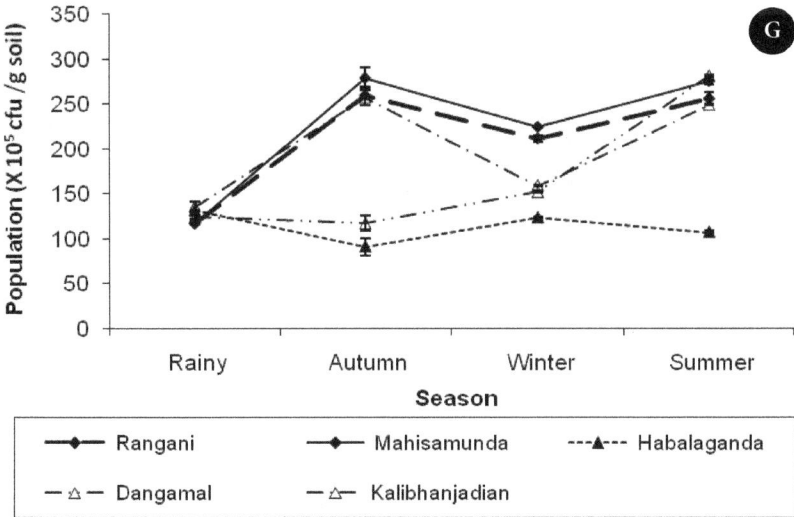

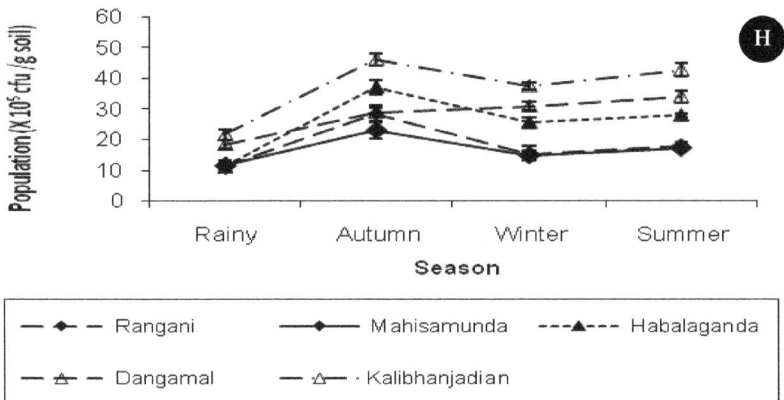

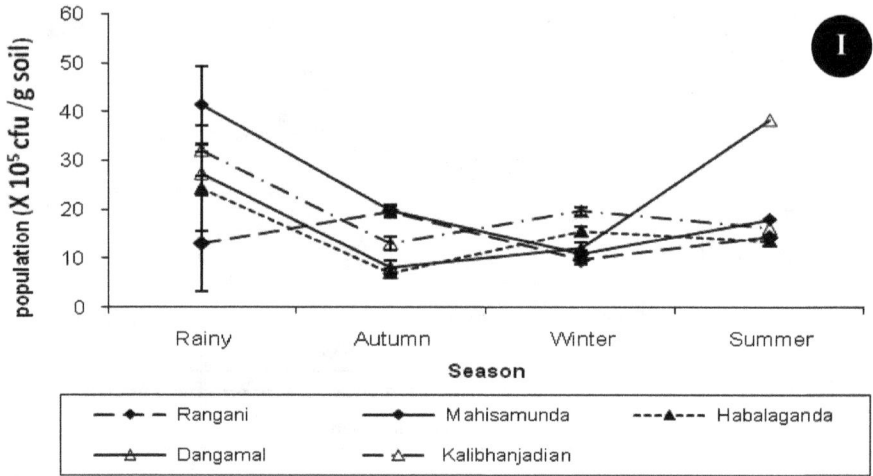

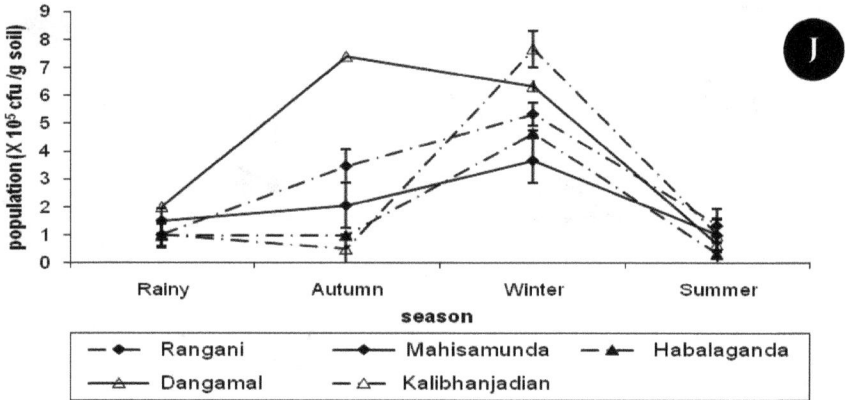

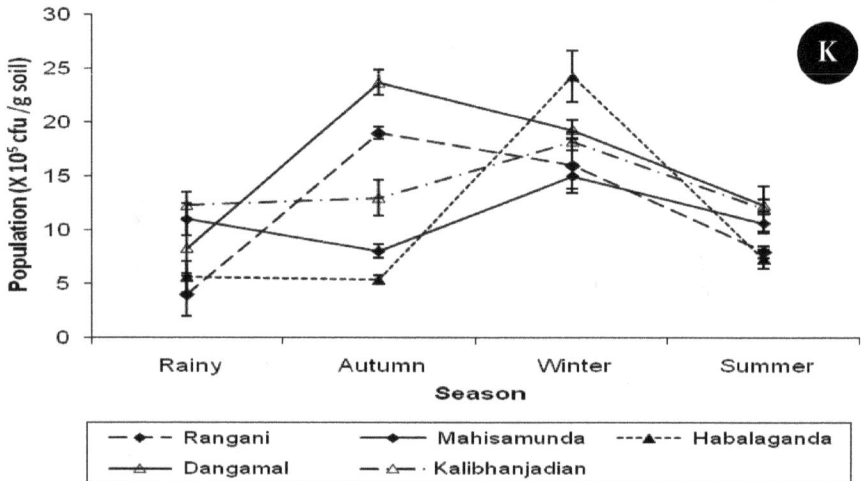

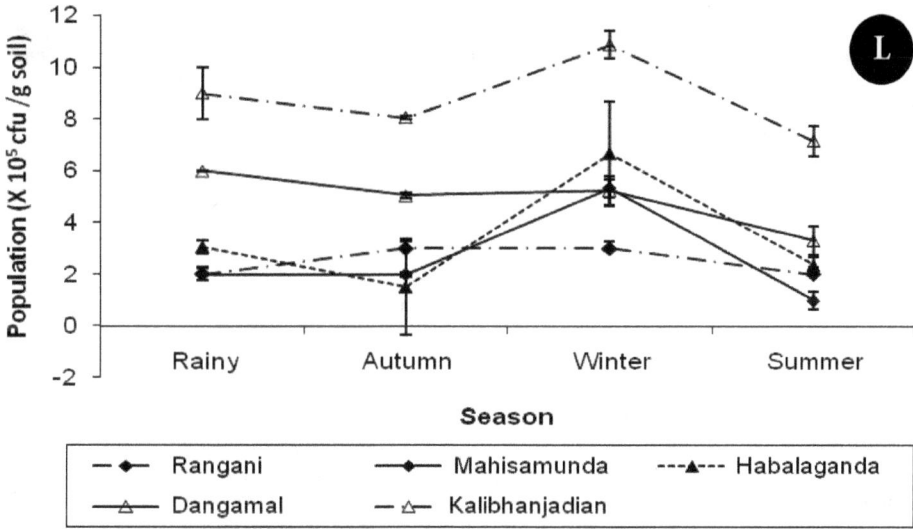

Fig: 21. Seasonal variation of bacterial population at five different sites of mangrove ecosystem of Bhitarkanika, Odisha, (A) heterotrophic, (B) Gram (-)ve, (C) Gram (+)ve, (D) P-solubilizing, (E) denitrifying, (F) nitrifying, (G) N2-fixing, (H) spore froming, (I) Sulphur oxidizing (J) Actinomycetes, (K) Anaerobic (L) Cellulose degrading bacteria

3.6. Principal Component Analysis

Principal component analysis (PCA), correlation and cluster analysis of population of different groups of bacteria and fungi and seven physico-chemical parameters of the soil were studied (Fig. 19). The data matrix consisting of 19 parameters were standardized through z-scale transformation in order to avoid misclassification due to dimensionally wide differences of data. The result of principal component analysis for five principal components (PCs) with eigen values greater than 1, and their significance towards the microbial association with the environment is presented in Fig. 22. In total 5 PCs represented 70.935% of cumulative variance which is sufficient to describe the relationship among the variables.

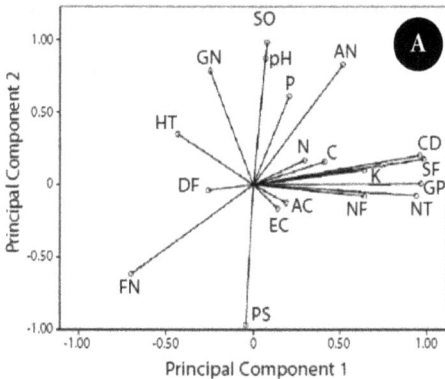

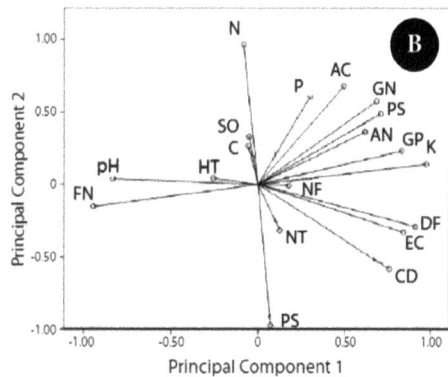

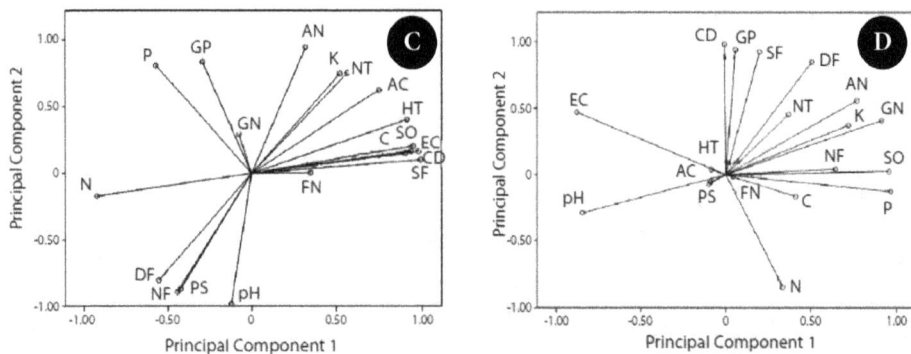

Fig: 22(a-d). Principal component analysis (PC1 vs PC2) of proximate variable.
Abbreviations: PS-phosphate solubilising, NT-Nitrifying, NF-N₂ fixing, CD-cellulose
degrading, DF-Denitrifying, EC-Electrical conductivity, N-Total Nitrogen, K-Total
Phosphorus, C-Total carbon, SF-Spore forming, AC-Actinomycetes, GP-Gram
positive, GN-Gram negative, AN-Anaerobic, P-Total phosphorus, FN-Fungus, HT-
Heterotrophic bacteria, (Rainy-A, Autumn-B, Winter-C, Summer-D.)

3.7. Phenotypic Identification of Four Important Groups of Bacteria from Mangroves of Bhitarkanika.

Some important groups of bacteria such as N_2 fixing, Phospahte solubilizing, Cellulose degrading and Sulphur oxidizing bacteria were isolated from mangrove soil using selective media and were subjected to various biochemical tests with a view to identify them. Results of some biochemical tests for phenotypic identification of the bacteria are given in Table 14-17 and Fig. 23

3.7.1. N₂ fixing bacteria

Fresh sediment samples were immediately processed for isolation of N_2 fixing microbes. The free living nitrogen fixing bacteria were isolated in a selective medium, comprising mannitol (15.0 g), K_2HPO_4 (0.5 g), $MgSO_4$. $7H_2O$ (0.2 g), $CaSO_4$ (0.1 g), NaCl (0.2 g), $CaCO_3$ (5.0 g), agar (15.0 g), distilled water (1 L) and pH 8.3. Pour-plating techniques were used for isolation of microbes from sediment samples. One gram fresh sediment was added to 9 ml water diluted serially through up to 10^{-5} and 0.1 ml of the soil suspension was mixed separately with different media and plated. The colony forming units (cfu) were counted after 3 days. Selected bacterial strains were subjected to biochemical characterization (Fig. 23) for their phenotypic identification (Table 14).

Table 14: Phenotypic characterization of free living N₂- fixing bacteria from mangroves of Bhitarkanika, Odisha

Characteristics	BNF1	BNF2	BNF3	BNF4	BNF5	BNF6
Shape	Rod	Rod	Rod	Rod	Rod	Rod
Cell diameter (μm)	1.4- 2.0	1.5- 2.5	2.5- 4.5	1.0-1.5	1.5-2.4	1.5-2.5
Gram stain	-	-	+	-	+	-
Motility	-	+	+	+	-	-

Characteristics	BNF1	BNF2	BNF3	BNF4	BNF5	BNF6
Spore	-	-	+	-	+	-
Spore round	-	-	-	-	-	-
Catalase test	+	+	+	+	+	+
Oxidase test	-	+	+	+	-	-
Indole test	-	+	-	-	-	+
Citrate test	+	+	-	+	-	-
MR test	-	+	+	+	+	+
MRVP test	+	-	-	-	-	-
Urease test	+	-	-	-	-	-
Nitrate reduction	+	-	-	+	+	-
Acid from:						
D- Glucose	+	+	+	+	+	+
Gas production:						
D-Glucose	+	-	-	+	-	-
D-Fructose	-	-	-	-	-	-
Sucrose	-	-	-	-	-	-
Hydrolysis of:						
Casein	-	-	+	-	-	-
Gelatine	-	-	+	-	-	-
Strach	+	+	+	+	+	+
Anaerobic growth	-	-	-	-	-	-
35°C	+	+	+	+	+	+
Arginine dihydrolase	-	-	-	-	-	+
H$_2$S production	-	-	-	-	-	-

BNF1: *Klebsiella* sp., **BNF2:** *Azotobacter* sp., **BNF3:** *Bacillus alcalophilus,* **BNF4:** *Pseudomonas* sp., BNF5: *Bacillus* sp., BNF6: *Pseudomonas putida* (BNF: Bhitarkanika N$_2$ fixing bacteria).

3.7.2. Phosphate solubilizing bacteria

Phosphate solubilising bacteria were isolated from soil samples. The soil samples were homogenized in sterile Milli Q water containing 0.85% NaCl (w/v), serially diluted (10^{-4}) and spreaded on Pikovskaya's agar medium plates. Then the petriplates were incubated at 30 °C for 24-48 h. Colonies were selected from the plates on the basis of the appearance of a clear halo zone. Once isolated, the bacterial isolates were maintained in the laboratory and were used for further study. Selected bacterial strains were subjected to biochemical characterization for their phenotypic identification. The species identified were *Klebsiella* sp., *Azotobacter* sp., *Bacillus alcalophilus, Pseudomonas* sp., *Bacillus* sp. and *pseudomonas putida* (Table 15).

Table 15: Phenotypic characterization of Phospahte solubilizing bacteria from mangroves of Bhitarkanika

Characters	PSB1	PSB2	PSB3	PSB4	PSB5	PSB6	PSB7	PSB8	PSB9
Shape	Rod	Rod	Rod	Rod	Rod	Rod	Rod	Rod	Rod
Cell diameter(µm)	1.0-1.5	1.55-1.65	1.58-1.82	1.76-1.88	2.05-1.52	1.55-1.98	1.59-.01	1.5-1.59	0.8-1.45
Spore	-	-	-	+	+	+	-	-	+
Motile	+	+	+	+	+	-	+	-	+
Aerobic	+	+	+	+	+	+	+	+	+
Gram stain	-	-	-	+	+	+	-	-	+
PHB	+	+	+	-	-	-	-	-	-
Catalase	+	+	+	+	+	+	+	+	+
Protease:									
Gelatin	-	+	-	+	-	-	-	-	-
Casein	-	-	-	+	-	-	-	-	-
Lipase:									
Tributyrin	+	+	+	+	-	-	-	-	-
Tween 80	+	-	+	+	+	-	-	-	-
Chitinase	-	-	-	+	+	-	-	-	-
Argine dihydrolase	-	+	+	+	+	-	+	+	+
Strach hydrolysis	-	-	-	-	+	-	+	-	-
Oxidase	+	-	+	+	+	+	-	+	+
Nitrate reduction	+	+	-	+	-	+	-	-	-
Acid from:									

Characters	PSB1	PSB2	PSB3	PSB4	PSB5	PSB6	PSB7	PSB8	PSB9
Shape	Rod	Rod	Rod	Rod	Rod	Rod	Rod	Rod	Rod
Cell diameter(µm)	1.0-1.5	1.55-1.65	1.58-1.82	1.76-1.88	2.05-1.52	1.55-1.98	1.59-.01	1.5-1.59	0.8-1.45
Glucose	+	+	+	+	+	-	+	+	+
Fructose	+	+	+	+	+	-	+	+	+
Mannose	+	+	+	+	+	-	+	+	+
Utilization of:									
Glucose	+	+	-	-	-	-	-	+	+
Arginine	-	+	+	+	+	-	+	+	+
Growth at									

PSB 1: *Pseudomonas cepacia*, PSB 2: *Pseudomonas sp.*, PSB3: *Pseudomonas stutzeri*, PSB4: *Bacillus sp.*, PSB5: *Bacillus sp.*, PSB6: *Bacillus lichiniformis*, PSB6: *Bacillus schlegelii*, PSB7: *Pseudomonas sp*, PSB- Phosphate solubilizing bacteria.

Table 16: Phenotypic characterization of cellulose degrading bacteria (CDB) from mangrove soils of Bhitarkanika, Odisha

Characteristics	CDB1	CDB2	CDB3	CDB4	CDB5	CDB6	CDB7
Shape	Rod	Rod	Rod	Rod	Rod	Rod	Rod
Cell diameter (µm)	0.15-0.17	1.0-1.45	1.0-1.55	1.5-1.75	1.0-1.25	1.00-1.55	1.65-1.75
Gram stain	+	-	+	+	+	-	-
Motility	-	-	-	+	+	-	-
Spore	+	-	+	+	+	-	-
Spore shape	Spherical	Elliptical	Spherical	Elliptical	Spherical	Elliptical	Elliptical
Spore position	Terminal	Central	Central	Central	Central	Central	Terminal

Characteristics	CDB1	CDB2	CDB3	CDB4	CDB5	CDB6	CDB7
Shape	Rod	Rod	Rod	Rod	Rod	Rod	Rod
Cell diameter (µm)	0.15-0.17	1.0-1.45	1.0-1.55	1.5-1.75	1.0-1.25	1.00-1.55	1.65-1.75
Catalase test	+	+	+	+	+	+	+
Indole production	-	-	-	-	-	-	-
VP test	+	-	+	+	-	-	+
Acid from							
D-Glucose	+	+	+	+	-	+	+
L- Arabinose	+	+	+	-	+	+	+
Gas from Glucose	-	-	-	-	-	-	-
Hydrolysis of							
Casein	-	-	+	+	+	-	-
Gelatin	-	-	+	+	+	+	-
Strach	+	+	+	+	+	+	+
Utilization of:							
Citrate	+	+	+	+	+	-	-
Nitrate reduction	-	+	+	+	+	+	+
Lecithinase	+	+	-	-	-	-	-
Anaerobic growth	-	-	-	+	-	-	-

CDB1: *Pseudomonas* **sp., CDB2**: *Pseudomonas* **sp., CDB3**: *Bacillus polymyxa*, **CDB4**: *Bacillus mycoides*, **CD5**: *Bacillus brevis*, **CD6**: *Pseudomonas* **sp., CDB7**: *Pseudomonas* sp, **CDB**: Cellulose degrading bacteria

3.7.3. Cellulose degrading bacteria

Cellulose degrading bacteria were isolated from soil samples using pour plate and spread plate techniques using CMC agar medium containing (Ray *et al.*, 2007): (g/l): Carboxymethylcellulose (CMC), 10; Tryptone, 2; KH_2PO_4, 4; Na_2HPO_4, 4; $MgSO_4.7H_2O$, 0.2; $CaCl_2.2H_2O$, 0.001; $FeSO_4.7H_2O$, 0.004; Agar,15 and pH adjusted to 7.0. The plates were incubated at 37 ^{0}C for 24-48 h. To visualize the hydrolysis zone, the plates were flooded with 0.1% Congo red solution and washed with 1M NaCl. Cellulose-producing activities of the isolates were estimated by the carboxymethyl cellulose hydrolysis capacity (HC value) on the cellulose congored agar, i.e. ratio of diameter of clearing zone and colony following the method of Lu *et al.*, (2005) and those with high HC values were selected and stored on slants at 4° C for the further study. Selected bacterial strains were subjected to biochemical characterization for their phenotypic identification. Based on morphological and biochemical analysis data 7 bacterial species were identified. The species identified are *Pseudomonas cepacia, Pseudomonas stutzeri, Pseudomonas* sp., *Bacillus* sp., *Bacillus lichiniformis, Bacillus schlegelii and Pseudomonas* sp. (Table 16).

3.7.4. Sulphur oxidizing bacteria

Sulphur oxidising bacteria were isolated from soil samples using pore plate method. The samples were poured on sulphur-oxidizer medium containing 10 g of peptone, 1.5 g of K_2HPO_4, 0.75 g of ferric ammonium citrate and 1.0 g of $Na_2S_2O_3.5H_2O$. The initial pH was adjusted to 7.0 using 1 M HCl before sterilizing by autoclave. Agar was added to a final concentration of 15 g per liter. The plates were incubated at 30 °C for 24 h. The morphologically distinct isolated colonies appeared on the plate were picked up by wire loop and re-streaked on the other sulphur-oxidizer medium agar plate for purity conformation. For qualitative screening of distinct sulphur oxidising bacteria, the isolated bacteria were further grown on the thiosulphate broth (Beijerinck, 1904) containing: 5.0 g $Na_2S_2O_3$, 0.1 g K_2HPO_4, 0.2 g $NaHCO_3$ and 0.1 g NH_4Cl in 1000 ml distilled water, with pH 8.0. Bromophenol blue was used as the indicator. The cultures which changed the colour of the thiosulphate broth from purple to colour less by reducing the pH after incubation for 3 days at 30 °C were selected for further characterization and identification. The species identified were *Desulfotomaculum* sp., *Desulfomonas* sp., *Desulfovibrio salexigens, Desulfovibrio* sp., *Desulfotomaculum* sp., *Pseudomonas* sp. and *Desulfotomaculum* sp. (Table 17).

Table 18: Phenotypic characterization of sulphur oxidizing bacteria (SOB) from mangrove soils of Bhitarkanika, Odisha

Characteristics	SOB1	SOB2	SOB3	SOB4	SOB5	SOB6	SOB7
Shape	Rod	Rod	Rod	Rod	Rod	Rod	Rod
Cell diameter (µm)	1.0-1.5	0.8- 1.2	0.5-1.0	1.0-1.5	1.0-1.5	1.5-2.5	1.5-2.5
Gram stain	-	-	-	-	-	-	-
Motility	-	-	-	-	-	-	-
Spore	-	-	-	-	-	-	-
Catalase test	-	-	-	-	+	+	+

Oxidase test	-	-	-	-	+	+	+
Indole test	-	-	+	-	-	-	-
Citrate utilization	-	-	-	-	+	+	-
MR test	+	+	+	+	+	+	+
MRVP test	+	-	+	-	+	+	-
Urease test	-	+	-	+	-	-	-
Nitrate reduction	-	-	-	-	+	+	+
Hydrolysis of:							
Casein	-	+	-	+	+	+	-
Gelatine	-	-	-	-	-	-	-
Strach	+	+	+	+	+	+	+
Anaerobic growth	-	-	-	-	-	-	-
Growth at 350C	+	+	+	+	+	+	+
Growth in:							
Lactate + sulfate	+	+	+	+	+	-	-
Pyruvate+sulfate	+	+	+	+	+	-	-
H2S production	+	+	+	+	+	+	+

SOB 1: *Desulfotomaculum* sp., **SOB2:** *Desulfomonas* sp., **SOB3:** *Desulfovibrio salexigens*, **SOB4:** *Desulfovibrio* sp., **SOB5:** *Desulfotomaculum* sp., **SOB6:** *Pseudomonas* sp., **SOB7:** *Desulfotomaculum* sp., **SOB-** Sulphur Oxidizing Bacteria

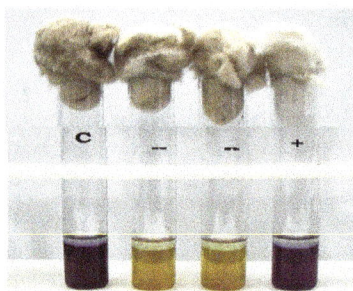

a. Arginine dihydrolase test

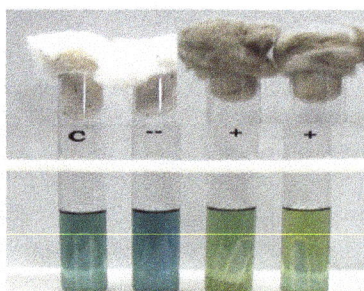

b. Acid-gas production test

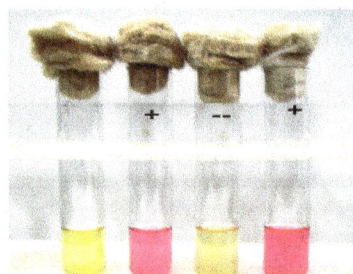

a. Urease test

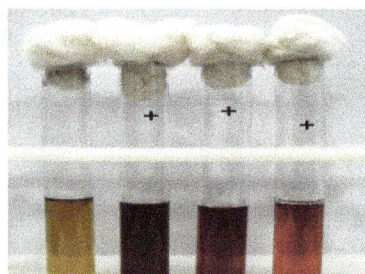

b. Voges Proskauer test

Continue....

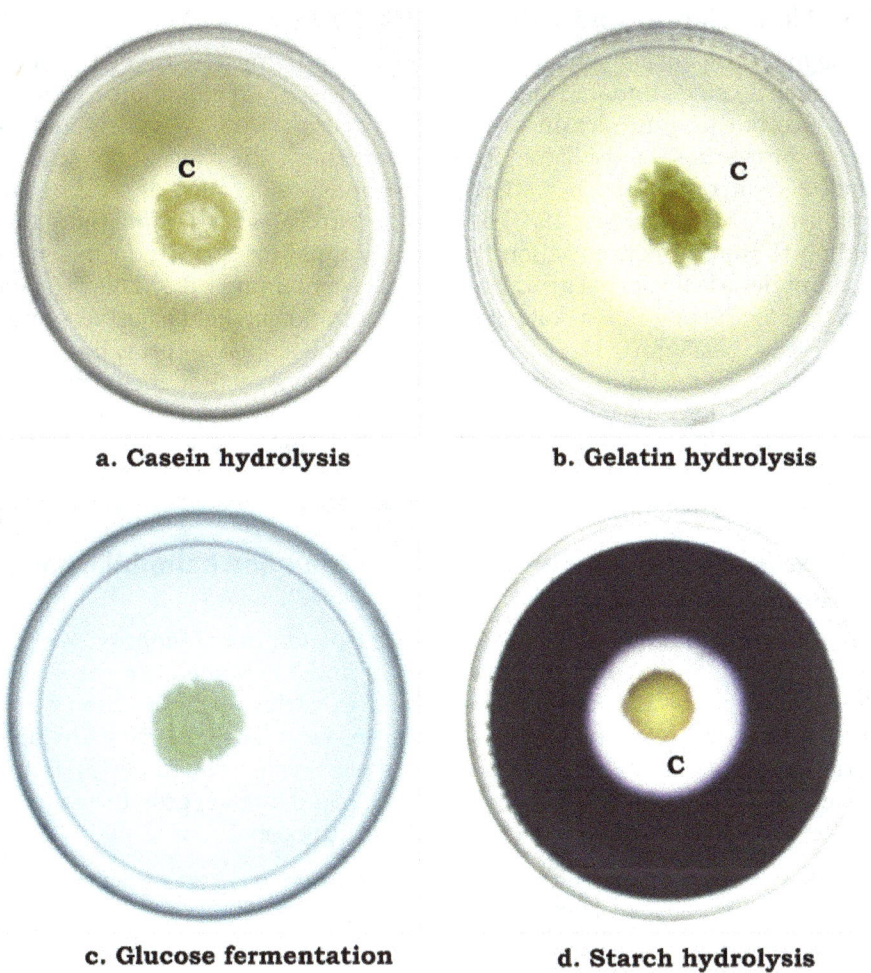

a. Casein hydrolysis   b. Gelatin hydrolysis

c. Glucose fermentation   d. Starch hydrolysis

Fig. 23: Some biochemical tests for identification of bacteria

3.8. Phenotypic and molecular studies on salt tolerant bacteria from Bhitarkanika mangrove soils

Since mangroves are saline environment, they harbour large number of salt tolerant bacteria (halotolerant and halophilic). Studies have been made to isolate and characterize some of the predominant halotolerant and halophilic bacteria from mangrove soil of Bhitarkanika with a view to assess their salt tolerant behaviour. The halotolerant and halophilic bacteria were isolated using enriched media. The halotolerant bacteria were isolated using nutrient agar media enriched with NaCl (7%). The halophilic bacteria were isolated using halophilic agar medium (MS97) and a detail investigation were carried out with respect to their morpho-physiological and molecular (genetic, proteomic and 16S rRNA gene sequencing) characterization for their identification.

3.8.1. Halotolerant and halophilic bacteria

The microorganisms, which are specialized for living in extreme hypersaline environments, as designated halophiles whereas those capable of growth in the absence of salt, but tolerant of varying concentrations, are considered to be halotolerant (Ventosa and Nieto, 1995). Nevertheless, the concepts of halophilic and halotolerant organisms, as well as their response to salt, vary depending on the criteria used. Thus, Kushner (1985) defined several categories of microorganisms according to the salt concentration that was optimal for growth. In this system, non-halophiles are those that grow best in media containing < 0.2 M NaCl (some of which, the halotolerant can tolerate higher concentrations), slight halophiles (marine bacteria) grow best with 0.2 to 0.5 M NaCl, moderate halophiles grow best with 0.5 to 2.5 M NaCl, and extreme halophiles show optimal growth in media containing 2.5 to 5.2 M NaCl (Ventosa and Nieto, 1995). One aspect of euryhaline halophiles which has received considerable attention involves their physiology, in particular, the physiological changes which occur as the organisms adapt to different salt concentrations (Vreeland, 1987).

3.8.2. Isolation and screening of halotolerant and halophilic bacteria in different growth conditions

Halotolerant and halophilic bacteria were isolated from mangrove soil using nutrient agar medium containing salt (NaCl) and MS97 medium respectively. Further, those bacterial strains were screened in various concentration of NaCl (5-15%) with a view to evaluate high salt tolerant bacteria. Similarly six halophilic bacteria were isolated and their salt tolerance was evaluated (Table-19). Both the halotolerant and halophilic bacteria were also screened for their growth in anaerobic condition as well as in different pH and temperature conditions (Table 18-19).

Table 18: Growth characteristics of halotolerant bacterial isolates from mangrove soils of Bhitarkanika, Odisha

Growth medium	BSB 7	BSB 8	BSB 9	BSB 10	BSB 11	BSB 13	BSB 14	BSB 15	BSB 16	BSB 17	BSB 18	BSB 19	BSB 20
NA	+	+	+	+	+	+	+	+	+	+	+	+	+
NA+NaCl(%):													
5	+	+	+	+	+	+	+	+	+	+	+	+	+
7	+	+	+	+	+	+	+	+	+	+	+	+	+
10	+	+	+	+	+	+	-	-	-	-	-	-	-
12	+	+	+	-	-	-	-	-	-	-	-	-	-
13	+	+	-	-	-	-	-	-	-	-	-	-	-
15	-	-	-	-	-	-	-	-	-	-	-	-	-
NA+Sea salt(%):													
5	+	+	+	+	+	+	+	+	+	+	+	+	+
10	+	+	+	+	+	+	+	+	+	+	+	-	+
15	-	-	-	-	-	-	-	-	-	-	-	-	-
NB+Sea salt(%):													
10	+	+	+	+	+	+	+	+	+	+	+	+	+

Growth medium	BSB 7	BSB 8	BSB 9	BSB 10	BSB 11	BSB 13	BSB 14	BSB 15	BSB 16	BSB 17	BSB 18	BSB 19	BSB 20
15	-	-	-	-	-	-	-	-	-	-	-	-	-
Anaerobic growth	-	-	-	+	+	-	-	-	-	-	-	-	-
Growth at temp:													
30 ºC	+	+	+	+	+	+	+	+	+	+	+	+	+
35 ºC	+	+	+	+	+	+	+	+	+	+	+	+	+
40 0C	+	+	+	+	+	+	+	+	+	+	+	+	+
Growth at pH:													
pH 3	-	-	-	-	-	-	-	-	-	-	-	-	-
pH 5	+	-	+	-	+	+	+	-	-	-	+	-	-
pH 7	+	+	+	+	+	+	+	+	+	+	+	+	+

BSB = Bhitarkanika Soil Bacteria, NA = Nutrient agar, NB = Nutrient broth, + = positive, - = negative

Table 19: Growth characteristics of halophilic bacterial isolates from mangrove soils of Bhitarkanika, Odisha grown in halophilic medium MS 97

Growth medium	Bacteria number					
	BSB 21	BSB 22	BSB 23	BSB 24	BSB 25	**BSB 26**
NA	+	+	+	-	-	+
NA + NaCl (%):						
10	+	+	+	+	+	+
12	-	-	-	+	+	+
13	-	-	-	+	+	+
15	-	-	-	+	+	+
20	-	-	-	+	+	+
25	-	-	-	+	+	-
NA+ Sea salt (%):						
10	+	+	+	+	+	+
15	-	-	-	+	+	+
NB+ Sea salt (%):						
10	+	+	+	+	+	+
15	-	-	-	+	+	+
Anaerobic growth in NB	-	-	-	+	+	-
Growth at temp:						
30 ºC	+	+	+	+	+	+
35 ºC	+	+	+	+	+	+

BSB = Bhitarkanika Soil Bacteria, NA = Nutrient agar, NB = Nutrient broth + = Positive, – = Negative

3.8.3. Molecular characterization and identification of bacteria

Out of the thirteen, six selected Gram (+) ve halotolerant bacterial strains (BSB1, 2, 3, 4, 6 and 12) were characterized by plasmid profile, genomic DNA and total cellular protein analysis. Out of these, finally two high salt tolerant bacterial strains (BSB 6 and BSB12) were selected for 16S rRNA gene sequencing. Similarly out of six halophilic bacterial strains one halophilic bacterial strain (H4) was subjected for 16s rRNA gene sequencing with a view to identify them. Based on blast search, homologues sequence alignment and phylogenetic tree construction the newly isolated and identified bacterial strains were placed in proper groups.

3.8.4. Genomic DNA, Plasmid and cellular protein profile of the isolates

The six Gram (+) ve bacteria (BSB 1, 2, 3, 4, 6, 12) produced the genomic DNA band of 23.59, 23.54, 21.79, 24.19 and 23.5 kbp, respectively (Fig. 24a). The organisms BSB1, BSB2, BSB4, BSB6 and BSB12 produced one plasmid each of 23.50, 26.08, 24.93, 25.21, and 23.49 kbp respectively (Fig. 24b). The strain BSB3 photograph was not presentable hence the data was not presented here. SDS-PAGE of the total cellular proteins of the isolates (BSB1, 2, 3, 4, 6 and 12) is given in Fig. 10c. The strain BSB1 produced 7 proteins of 114, 95, 66, 16, 13, 12 KDa, BSB2 produced 6 proteins of 113, 94, 64, 17.2, 13, 11.5 KDa, BSB3 produced 5 proteins of 114, 93, 66.2, 37.4, 35 KDa, BSB4 produced 4 proteins of 114, 93, 66.2, 37.4 KDa, BSB 6 produced only one protein of 66.2 KDa while BSB12 produced 4 proteins of 114, 93, 66.2, 37.4 KDa respectively (Fig. 24c).

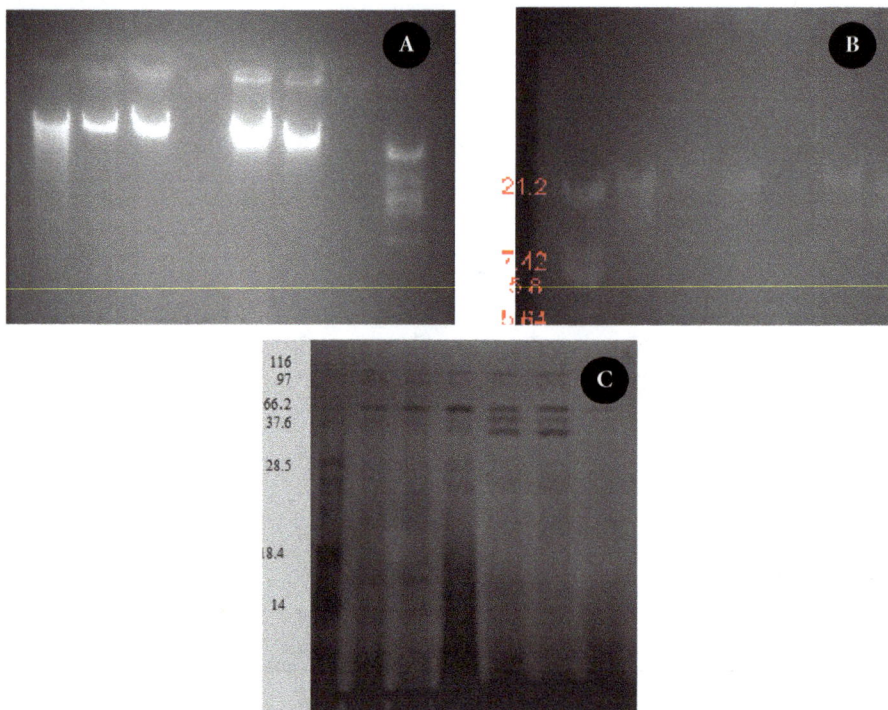

Fig 24(a): Genomic DNA, (b) plasmid and (c) Protein profile of the bacterial strains

3.8.5. 16S rRNA gene sequencing of two bacterial isolates

The 16S rRNA gene sequencing was carried out for two halotolerant bacterial strains (BSB6 and BSB12) and one halophilic bacterial strain (H4). The genomic DNA was isolated from the selected strains then subsequently amplified using 16S rRNA gene specific primer. Then the amplicons was subjected for sequencing using automated sequencer machine. The boot strap and phylogenetic analysis revealed that BSB12 matched 100% with the *Bacillus megaterium*. BSB12 runs in parallel evolution with *B. megaterium* while BSB6 stood separately sharing 97% homology with *B. megaterium*. The phylogenetic tree constructed based on 16S rRNA gene sequences was shown in Fig. 25. Based on 16S rRNA gene sequencing the two isolates (BSB6 and BSB 12) were identified as *Bacillus megaterium* (Fig. 25a). The strains, BSB6 and BSB12, were deposited in the Microbial Type Culture Collection Centre; Chandigarh (India) with MTCC numbers MTCC-9204 and MTCC-9205 respectively. The 16S rDNA sequences of the strains were submitted to Gene Bank with accession numbers FJ853653 and FJ973576. The 16S rRNA gene sequencing and phylogenetic analyses of the strain H4 showed 97% homology with *Vigribacillus* sp. (MTCC 9880) and the same was submitted to Gen Bank with accession number JQ945737 (Fig. 25b).

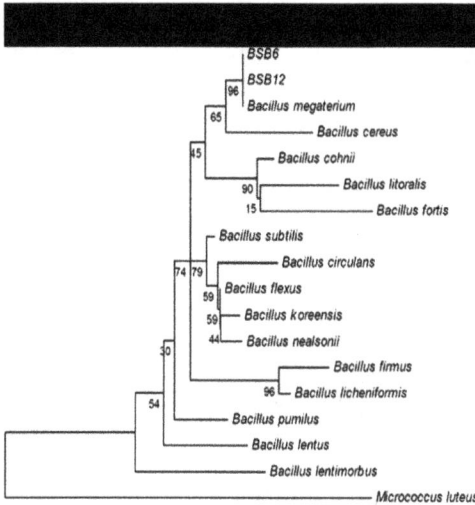

Fig 25(a) *Bacillus megaterium*   Fig. 25(b) *Vigribacillus* sp.

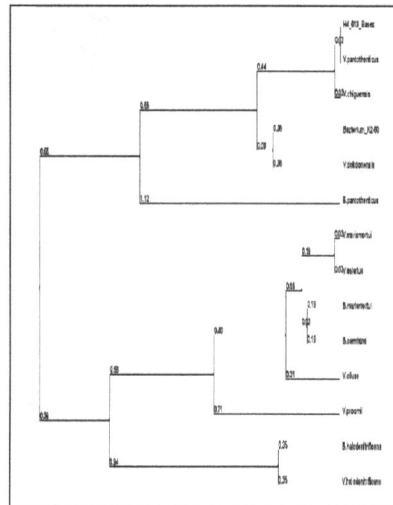

Fig 25 (a, b): 16S rRNA gene sequence based neighbor-joining tree showing the position of strain BSB 6, BSB 12 and H4. Numbers at node indicate bootstrap values. In the boot strap a multiple alignment is re-sampled 500 times.

3.8.6. SEM study of selected bacterial strains

Besides sequencing, the Scanning Electron Microscope (SEM) photos of few selected bacterial strains were taken. For SEM study, the bacterial cell suspension was centrifuged at 10,000g for 10 min and the pellet was washed with Tris–HCl buffer followed by deionized water three times. The samples were dehydrated with

70% ethanol, mounted on an aluminium stub, coated with gold and examined under JEOL (840A, Japan) SEM at 200 kV. From the SEM study it is revealed that most of the bacterial strains are rod shaped with endospores (Fig. 26).

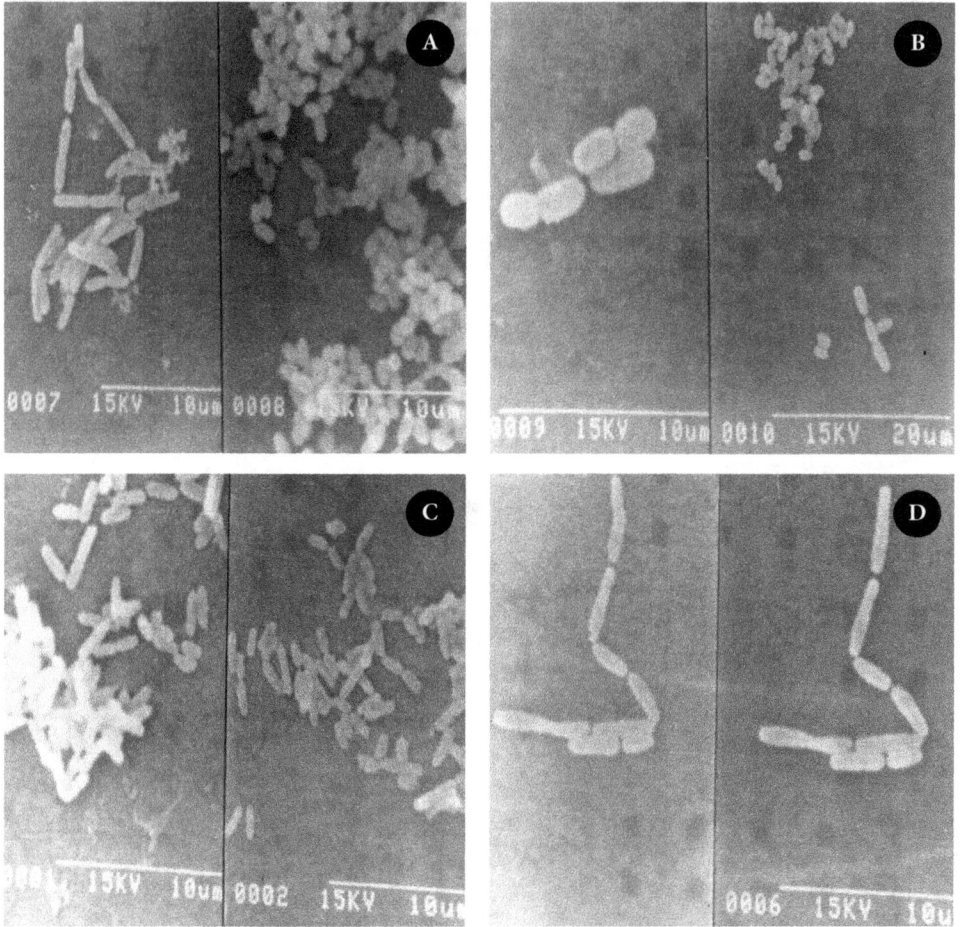

Fig: 26 Scanning Electron Microscope (SEM) photographs of some mangrove bacterial isolates A, B - *Bacillus* sp., D - *Pseudomonas* sp.

3.9. Ennumeration of Different Groups of Bacteria from Mangroves of Mahanadi Delta

Different groups of bacterial population such as heterotrophic, Gram (+)ve, Gram (-)ve, nitrogen fixing, nitrifying, denitrifying, spore forming, phospahte solubilizing, cellulose degrading and actinomycetes were enumerated from soil samples collected from six different locations of mangroves of Mahanadi delta during four different seasons (rainy, autumn, winter and summer) and is presented in Fig, 27a- Among the different groups of bacteria studied, the heterotrophic bacterial population (46-167 x 10^5cfu/g soil) was observed to be maximum during rainy season (Fig. 27a). The free living N_2 fixing bacterial population (28-113 x

10^5cfu/g soil) was found to be fluctuated alternatively during the study period, which however was maximum during winter and minimum during the rainy season (Fig. 27b). The Gram (-)ve bacterial population varied between 8- 134 x 10^5 cfu/g of soil in different sites, which increased 2 fold at Kharnasi than the other sites during the rainy season (Fig. 27c). The nitrifying bacterial population showed a clear seasonal variation (8.33-104 x 10^5 cfu/g soil) which is less in summer season than the rainy season (Fig. 27d). The sulphur oxidising bacterial population was nearly same in autumn and winter but gradually increased towards summer (5.6-78) (Fig. 27e). The population of Gram (+) ve bacteria varied from 9.33-110 (x 10^5 cfu/g soil) in different sites and season with a peak during the winter season (Fig. 27f). The spore forming bacterial population followed a common trend which was more in the summer season (5.3 -60 x 10^5 cfu/g soil) and declined steadily towards rainy season (Fig. 27g). The population of denitrifying bacteria varied from 4.33-114 (x 10^5 cfu/g. soil), showing highest peak during winter season (Fig. 27h). The population of anaerobic bacteria varied from 2-12 (x 10^5 cfu/g soil), which is found to be less in rainy and increased towards winter and summer (Fig. 27i). The phosphate solubilising bacteria (1.00-7.00 x 10^5 cfu/g soils) recorded with maximum population during the rainy season (Fig. 27j). The cellulose degrading bacterial population varied from 1.33-34 (x 10^5 cfu/g soil) and showed a maximum value in winter season (Fig. 27k).

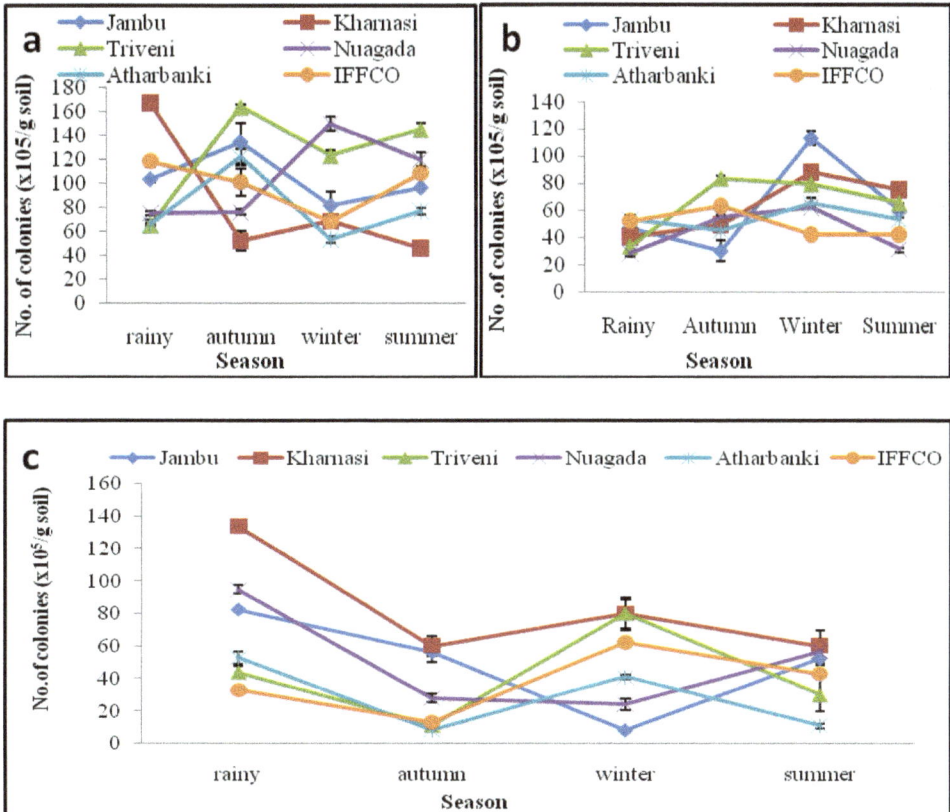

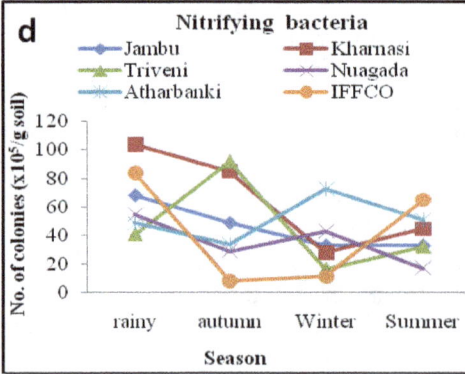

d Nitrifying bacteria
Jambu, Kharnasi, Triveni, Nuagada, Atharbanki, IFFCO

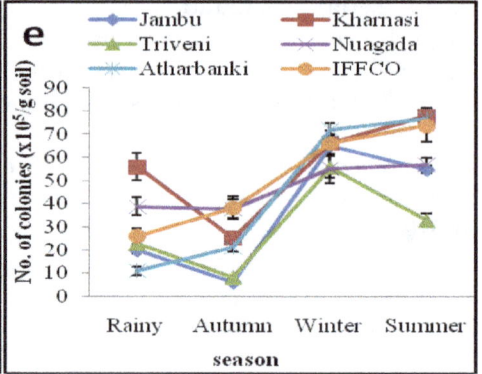

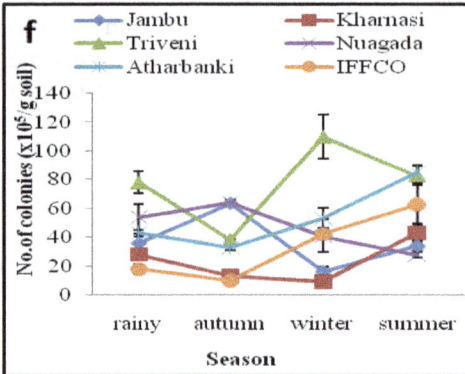

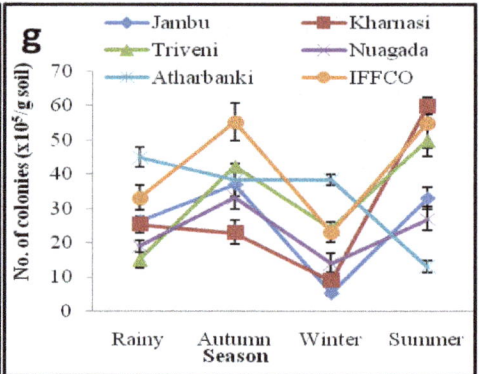

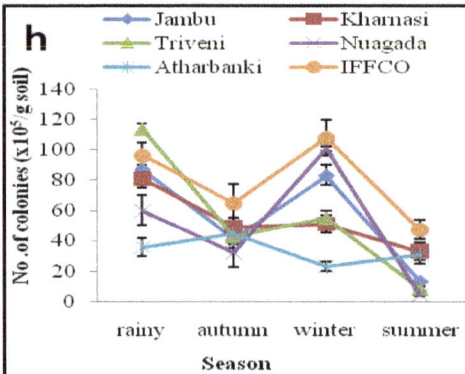

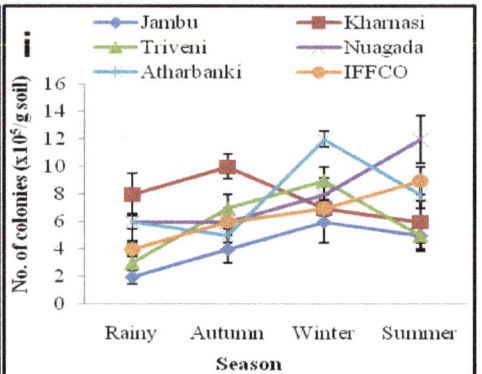

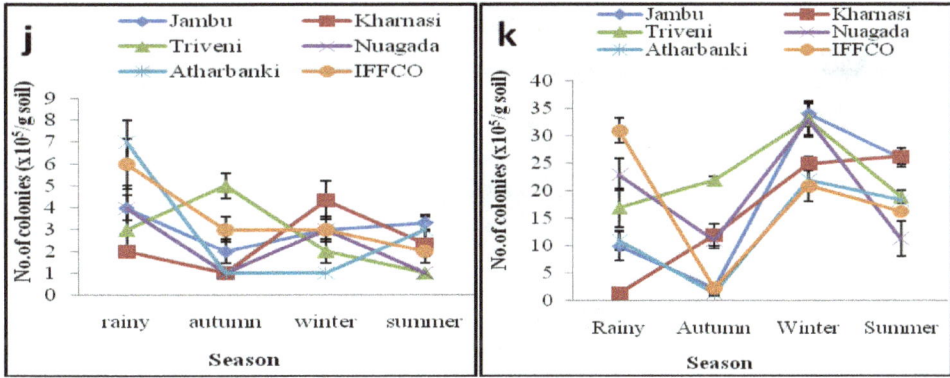

Fig. 27 (a-k) Seasonal variation of microbial population in mangrove soils of Mahanadi delta (a) Heterotrophic bacteria, (b)N$_2$ fixing bacteria, (c) Gram (–)ve bacteria, (d) Nitrifying bacteria, (e) Sulphur oxidising bacteria, (f) Gram (+ ve) bacteria, (g) Spore producing bacteria, (h) Denitrifying bacteria, (i) Anaerobic bacteria, (j) Phosphate solubilising bacteria, (k) Cellulose degrading bacteria

3.10. Principal Component Analysis

The principal component analyses (PCA) for the soil samples of the Mahanadi delta from six different sites during four seasons (summer, rainy, autumn and winter) are shown in Fig. 28 a-d. The PCA for the summer season was carried out and the result is presented in Fig. 25a. The four factors of the PCs explain 83.39 % of the total variance. PC-1 accounts for 26.05 % of the total variance with eigen value of 4.429, which is due to strong positive load of sulphur (0.874), phosphorous (0.859) along with nitrifying bacteria (0.855) and denitrifying bacteria (0.796) and strong negative load of pH (-0.703). The PC-2 accounts for 24.60 % of the total variance with eigen value of 4.183, which is due to strong positive load of sulphur (0.877), sulphur oxidizing bacteria (0.831) and phosphate solubilising bacteria (0.761) and strong negative load of heterotrophic bacteria (-0.863). The PC-3 accounts for 20.44 % of the total variance with eigen value of 3.476, which is due to strong positive load of cellulose degrading bacteria (0.810), nitrogen fixing bacteria (0.888) and a negative load of anaerobic bacteria (-0.890). Likewise the PC-4 accounts for 12.28 % of the total variance. Further, the correlation analysis of soil nutrient content and bacterial population in different sites during the summer season showed a significant positive correlations between carbon content and denitrifying bacteria (r=0.724) as well as salinity and phosphorous content (0.803). Due to higher nutrient content (nitrate and phosphate), there is increase in cellulose degrading bacteria, sulphur oxidizing bacteria, denitrifying bacteria etc. Cellulose degrading bacteria is positively correlated with nitrate content (r=0.561) and potassium content (r=0.594).

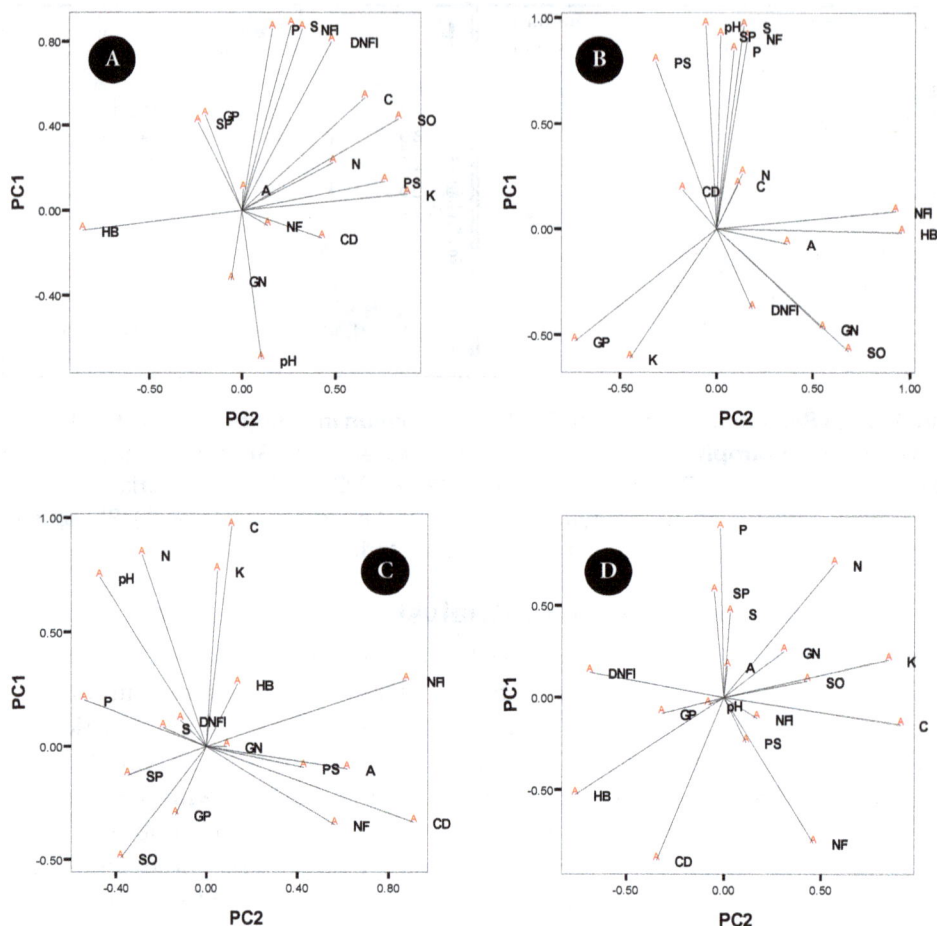

Fig. 28 Principal Component analysis between soil population and soil nutrient content in summer (a), rainy (b), autumn (c) and winter (d)

3.11. Phenotypic Identification of Some Major Groups of Bacteria from Mangroves of Mahanadi Delta

3.11.1. Identification of Phospahte solubilizing bacteria

Phospahte solubilizing bacteria were isolated from mangrove soil samples using Pikovasky agar medium. The bacterial strains those produced halo zone (Fig. 29) in the Pikovasky agar medium were considered as phosphate solubilizers. Further, in the basis of halo zone formation, 12 bacterial strains were subjected to various biochemical tests with a view to identify them (Table 20). Based on morphological and biochemical characterization, the strains were identified as *Bacillus* sp; *Pseudomonas* sp., *B. subtilis.*, *B. megaterium*, *Alcaligens fecalis.*, *Klebsiella* sp., *Serratia* sp., *Azotobacter* and *Micrococcus* sp.,

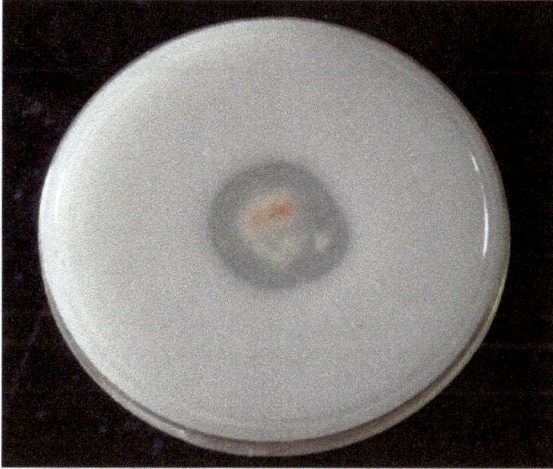

Fig. 29: Formation of halozones by 'P' solubilising bacteria on NBRIP-agar medium

3.12.2. Molecular Identification of phosphate solubilizing bacteria

Two highly phospahte solubilizing bacteria were subjected to 16SrRNA gene sequencing for conformation at species level. Further confirmation of genus *Alcaligenes* was done up to species level by BLAST analysis of the 16S rRNA gene sequence that showed 100% similarity with genus *Alcaligenes*. A phylogenetic tree was constructed by comparing nucleotide sequences of 16S rRNA gene sequences of PSB-26 with different *Alcaligenes* sp. submitted in NCBI database and found that the isolate is closest to *Alcaligenes fecalis subsp. phenolicus* and PSB-37 as *Serratia* sp. (Fig. 30). Scanning Electron Microscope photo graphs of these two bacterial strains (PSB 26 and PSB37) reveled that both are rod shaped with centrally located endospores (Fig. 31).

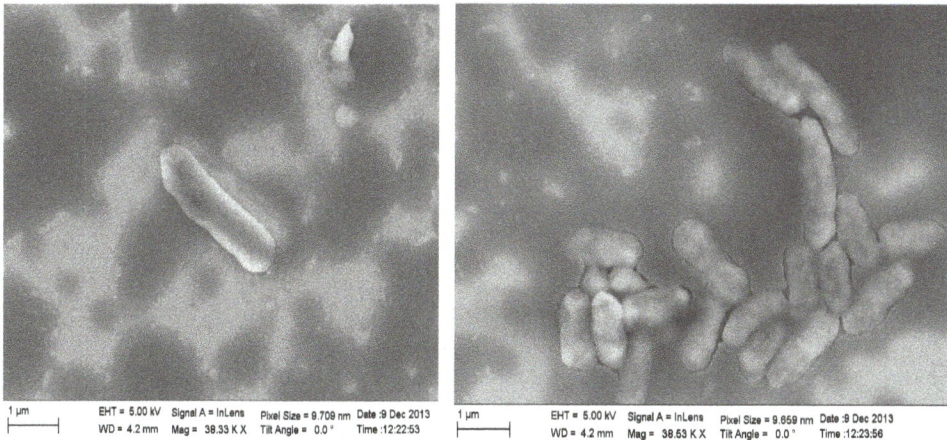

Fig. 31. SEM photograph of PSB-26 and (b) PSB-37

Table: 20 Morphological and biochemical characterization of phosphate solubilizing bacteria isolated from mangrove of Mahanadi delta

Characters	PSB16	PSB18	PSB21	PSB26	PSB27	PSB28	PSB29	PSB34	PSB37	PSB41	PSB44	PSB48
Shape	Rod	Rod	Rod	Rod	Rod	Rod	Rod	Rod	Rod	Rod	cocci	cocci
Cell diameter(µm)	0.8-0.9	0.76-0.88	1.0-1.05	0.55-0.95	0.9-1.0	0.75-0.9	0.8-1.45	0.4-0.8	0.7-2.3	1.55-1.98	1.5-2.0	0.9-2.1
Spore	-	+	+	-	-	-	+	-	-	-	-	-
Motility test	+	+	+	+	+	+	+	+	+	-	+	-
Aerobic growth	+	+	+	+	+	+	+	+	+	+	+	+
Urease production	-	-	+	-	+	-	-	+	+	+	-	+
Gram stain	-	+	+	+	-	-	+	-	-	-	-	+
PHB	+	-	-	+	+	+	-	+	+	+	+	+
Catalase production	+	-	+	+	+	+	+	+	+	+	+	+
MR	-	+	+	-	+	-	-	-	-	-	+	-
VP	-	+	+	-	-	-	-	+	+	-	-	-
Citrate utilisation	-	+	+	+	+	-	+	+	+	+	+	+
Protease:												
Gelatin hydrolysis test	-	+	+	-	-	-	+	-	+	-	-	-
Casein hydrolysis test	-	+	+	-	-	-	+	-	+	-	-	-
Lipase:												
Tributyrinhydrolysis test	+	+	-	-	-	-	-	-	+	-	-	-
Tween 80 utilisation	+	+	+	-	-	-	-	+	+	-	-	-
Lecithinase egg yolk	-	-	-	-	+	-	-	-	-	+	-	-
Chitinase production	-	+	+	-	-	-	-	-	-	-	-	-
Arginine dihydrolase	+	+	+	-	+	+	+	-	-	-	-	-
Strach hydrolysis	-	-	+	-	+	-	+	+	-	-	+	-
Oxidase production	+	+	+	+	-	+	+	-	-	+	+	+
Nitrate reduction	-	-	+	-	-	-	-	+	+	+	+	+
Acid from:												
Glucose	+	+	+	-	+	+	+	+	+	-	+	-
Fructose	+	+	+	-	+	+	-	+	+	-	+	-

Characters	PSB16	PSB18	PSB21	PSB26	PSB27	PSB28	PSB29	PSB34	PSB37	PSB41	PSB44	PSB48
	Pseudomonas sp.	*B. pumilus*	*B. subtilus*	*Bacillus sp.*	*Pseudomonas sp.*	*Pseudomonas sp.*	*B. megaterium*	*Alcaligens sp.*	*Klebsiella sp*	*Serratia sp.*	*Azotobacter sp.*	*Micrococcus sp.*
Mannose	+	+	+	-	+	+	-	+	+	-	+	-
Gas production:												
Glucose	-	-	+	+	-	+	-	+	-	-	+	+
Indole	-	-	-	-	-	-	-	-	-	-	-	+
Growth pH 5-7	+	+	+	+	+	+	+	+	+	+	+	+
Growth 40° C	+	+	+	+	+	+	+	+	+	+	+	+

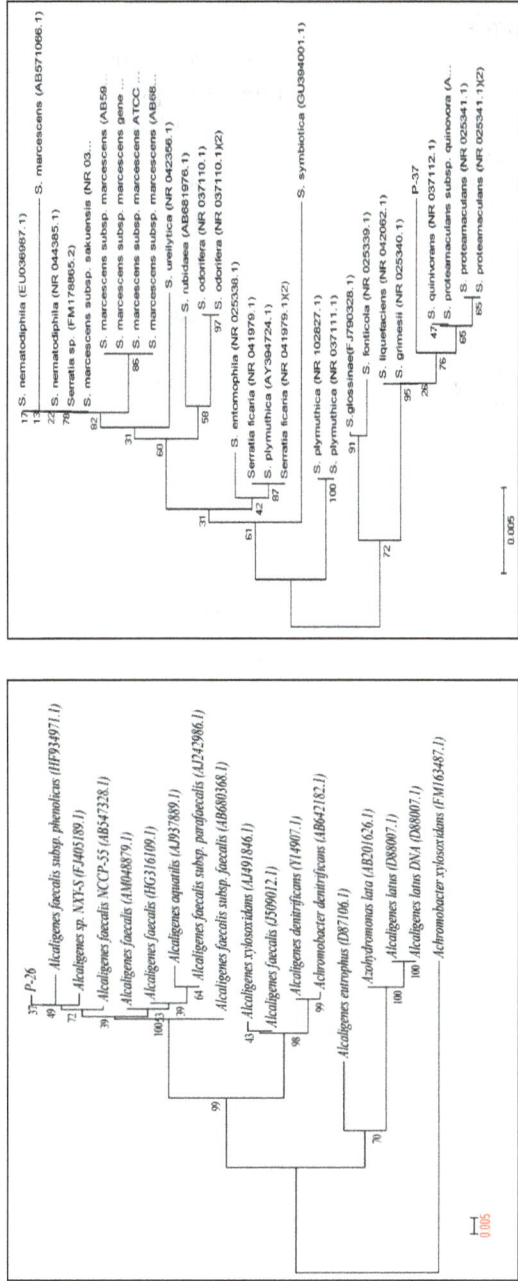

Fig. 30 : Phylogenetic tree of *Alcaligenes fecalis sub sp. phenolicus* (PSB-26) and *Serratia sp.* (PSB-37).

3.13. Sulphur oxidizing bacteria isolated from Mangrove of Mahanadi delta, Odisha

For isolation of sulphur oxidising bacteria, soil samples were collected from six different locations of mangroves of Mahanadi delta. Soil sample were taken, homogenized in sterile MilliQ water containing 0.85% NaCl (w/v) and serially diluted (10^{-4} times) and poured on sulphur-oxidizer medium containing 10 g of peptone, 1.5 g of K_2HPO_4, 0.75 g of ferric ammonium citrate and 1.0 g of $Na_2S_2O_3.5H_2O$. The initial pH was adjusted to 7.0 using 1 M HCl before sterilizing by autoclave. Agar was added to a final concentration of 15 g per liter. The plates were incubated at 30 °C for 24 h for growth. For qualitative screening, the isolated bacteria were further grown on the thiosulphate broth (pH 8.0) with bromophenol blue as the indicator. The cultures which changed the colour of the thiosulphate broth from purple to colour less by reducing the pH after incubation for 3 days at 30°C were selected for further characterization and evaluation of their sulphate ion and sulphide oxidase activity. Based on the morphological and biochemical analysis, SOB-7 tentatively belongs to genus *Klebsiella* sp. and SOB-8 belongs to *Micrococcus* sp. (Fig. 32). Further confirmation of genus *Klebsiella* sp. and *Micrococcus* sp were done by BLAST analysis which showed 100% similarity with genera *Klebsiella* sp. (SOB-7) and *Micrococcus* sp (SOB-8). A phylogenetic tree was constructed by comparing nucleotide sequences of 16S rRNA gene sequences of SOB-7 & SOB-8 with different *Klebsiella* sp. and *Micrococcus* sp, submitted in NCBI database and found that the isolate SOB-7 is closest to *Klebsiella* sp. (Fig. 33), where as SOB-8 to *Micrococcus* sp..

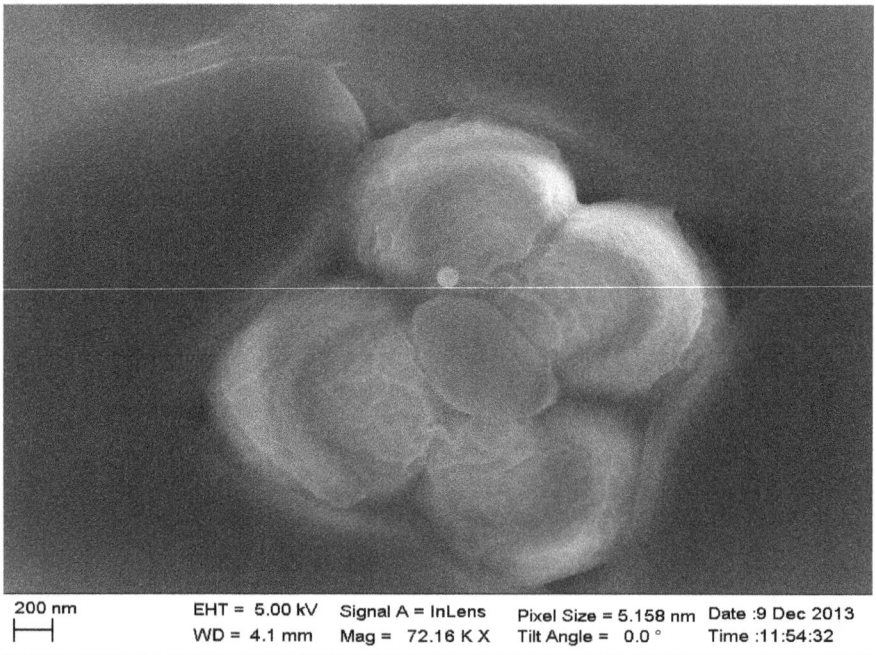

| 200 nm | EHT = 5.00 kV | Signal A = InLens | Pixel Size = 5.158 nm | Date :9 Dec 2013 |
| | WD = 4.1 mm | Mag = 72.16 K X | Tilt Angle = 0.0 ° | Time :11:54:32 |

Fig. 32: SEM photograph of *Microcuus* sp.

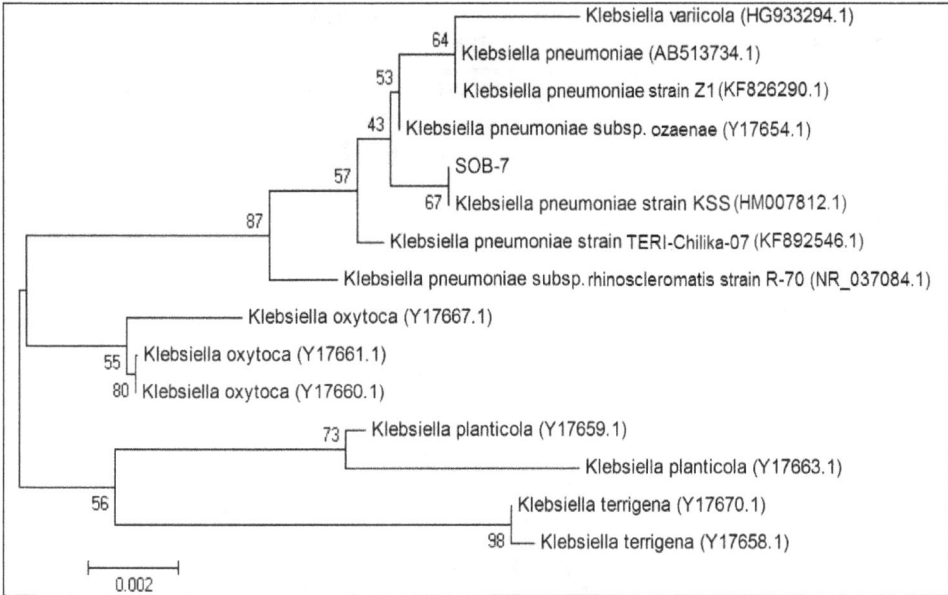

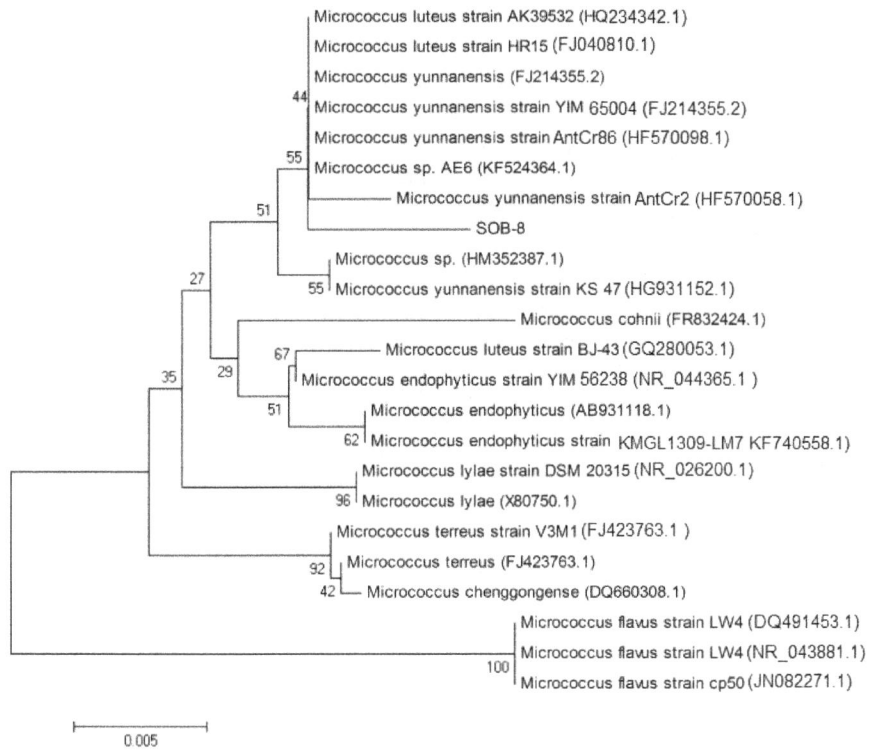

Fig. 33: Phylogenetic tree of *Klebsiella* sp. (SOB-7) and *Microccus* sp. (SOB-8) isolated from mangrove soil of Mahanadi river delta.

3.13.1. Isolation and identification of cellulose degrading bacteria isolated from mangrove soil of Mahanadi delta

The cellulose degrading bacteria were isolated from the soil samples collected from different location of mangrove ecosystem. Soil sample were taken, homogenized in sterile milli Q water containing 0.85% NaCl (w/v) and serially diluted to 10^{-4} times and pour plate and spread plate techniques were done using CMC agar medium containing (g/l): Carboxymethylcellulose (CMC), 10; Tryptone, 2; KH_2PO_4, 4; Na_2HPO_4, 4; $MgSO_4.7H_2O$, 0.2; $CaCl_2.2H_2O$, 0.001; $FeSO_4.7H_2O$, 0.004; Agar,15 and pH adjusted to 7.0. The plates were incubated at 37 °C for 24-48 h. To visualize the hydrolysis zone (**Fig. 34**), the plates were flooded with 0.1% Congo red solution and washed with 1M NaCl. Altogether, 15 cellulose degrading bacteria were isolated from mangrove soil and subjected to various biochemical tests with a view to identify them (Table 22).

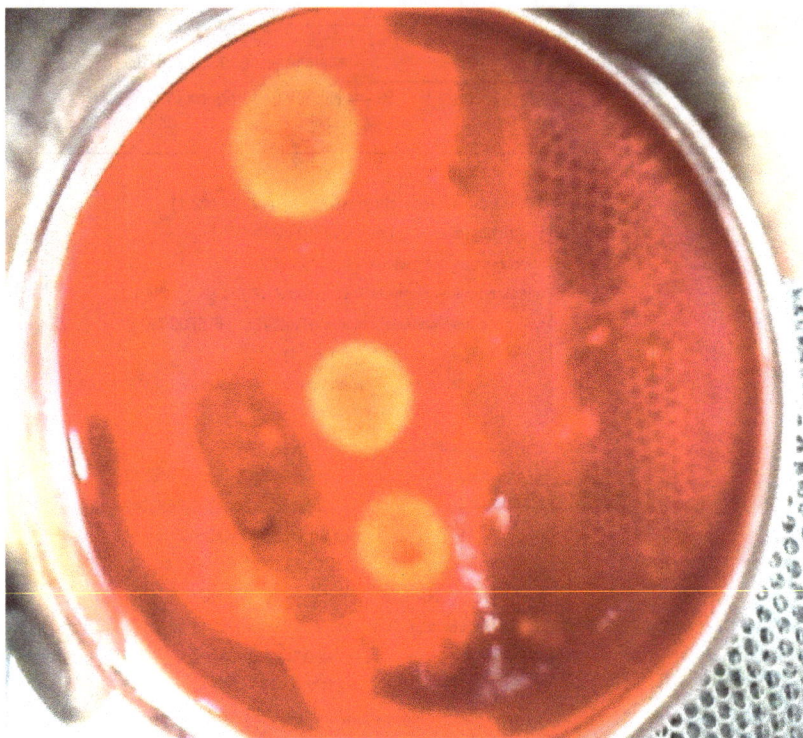

Fig. 34: Clearing zone generated by cellulose degrading bacteria in CM cellulose Congo red agar medium

Table 21: Phenotypic characterization of cellulose degrading bacteria from mangroves of Mohanadi delta.

Characters	CDB1	CDB2	CDB3	CDB4	CDB5	CDB7	CDB8	CDB9	CDB11	CDB12	CDB13	CDB14	CDB15
Shape	Rod	Rod	Rod	cocci	Rod	Rod	Rod	Rod	Rod	Rod	Rod	Rod	cocci
Cell diameter (µm)	0.8-0.9	0.7-0.8	0.8-0.1	0.8-2.0	0.5-0.7	0.9-0.1	0.5-0.7	1.0-1.05	0.9- 1.0	1.55-1.98	0.9- 1.0	1.55-1.98	1.0-2.1
Spore	+	+	-	-	-	-	-	+	-	+	-	+	-
Motility test	+	+	+	-	-	+	-	+	+	-	+	-	-
Aerobic growth	+	+	+	+	+	+	+	+	+	+	+	+	+
Urease	-	-	-	+	+	-	+	-	+	+	+	+	+
Gram staining	+	+	-	+	-	-	-	+	-	+	-	+	+
PHB	ND	ND	+	-	ND	+	ND	-	-	-	-	-	-
Catalase test	+	+	+	+	+	+	+	+	+	+	+	+	+
MR	-	-	-	-	-	-	-	+	+	-	+	-	-
VP	-	-	-	-	-	-	-	+	-	-	-	-	-
Citrate utilisation	+	-	-	+	+	-	+	+	+	+	+	+	+
Protease:													
Gelatin hydrolysis	+	+	+	-	+	-	+	+	-	-	-	-	-
Casein hydrolysis	+	+	+	-	+	+	+	+	-	-	-	-	-
Lipase:													
Tributyrin hydrolysis	ND	ND	ND	-	-	ND	-	-	-	-	-	-	-
Tween 80 hydrolysis	ND	ND	ND	-	+	ND	+	+	-	-	-	-	-

Characters	CDB1	CDB2	CDB3	CDB4	CDB5	CDB7	CDB8	CDB9	CDB11	CDB12	CDB13	CDB14	CDB15
	B. brevis	*B alcalophilus*	*Xanthomonas sp*	*Micrococcus sp.*	*Brucella sp.*	*Xanthomonas sp*	*Brucella sp.*	*B. subtilis*	*Pseudomonas sp.*	*Bacillus sp.*	*Pseudomonas sp*	*Bacillus sp.*	*Micrococcus sp.*
Lecithinase egg yolk	-	-	-	-	-	-	-	-	+	+	+	+	-
Chitinase production	-	-	-	-	-	-	-	+	-	-	-	-	-
Argine dihydrolase	-	-	-	-	-	-	-	+	+	-	+	-	-
Strach hydrolysis	+	+	+	-	+	-	+	+	+	-	+	-	-
Oxidase production	-	+	-	+	+	-	+	+	-	+	-	+	+
Nitrate reduction	+	-	-	+	+	-	+	+	-	+	-	+	+
Acid from:													
Glucose	ND	ND	+	-	-	+	-	+	+	-	+	-	-
Fructose	ND	ND	+	-	-	+	-	+	+	-	+	-	-
Mannose	ND	ND	+	-	-	+	-	+	+	-	+	-	-
Gas production	-	-	-	-	-	-	-	-	-	-	-	-	-
Indole	-	-	-	-	-	-	-	-	-	-	-	-	-

ND=Not Detected

3.13.2. Molecular identification of cellulose degrading bacteria

Based on the morphological and biochemical analysis showed, CDB-5 tentatively belongs to genus *Brucella* sp. and CDB-12 belongs to the *Bacillus* sp. (Table 5). Further confirmation of genus *Brucella* (CDB-5) and *Bacillus* (CDB-12) were done by BLAST analysis data of 16S rRNA gene sequence that showing 100% similarity with genera *Brucella* and *Bacillus* sp. respectively. Phylogenetic tree were constructed by comparing nucleotide sequences of 16S rRNA sequences of CDB-5 with different *Brucella* sp. and 16S rRNA sequences of CDB-12 with different *Bacillus* sp. respectively and submitted in NCBI database and confirmed that the isolate CDB-5 is closest to *Brucella* sp. (Fig. 35) and CDB-12 is closest to *Bacillus* sp. (Fig. 36).

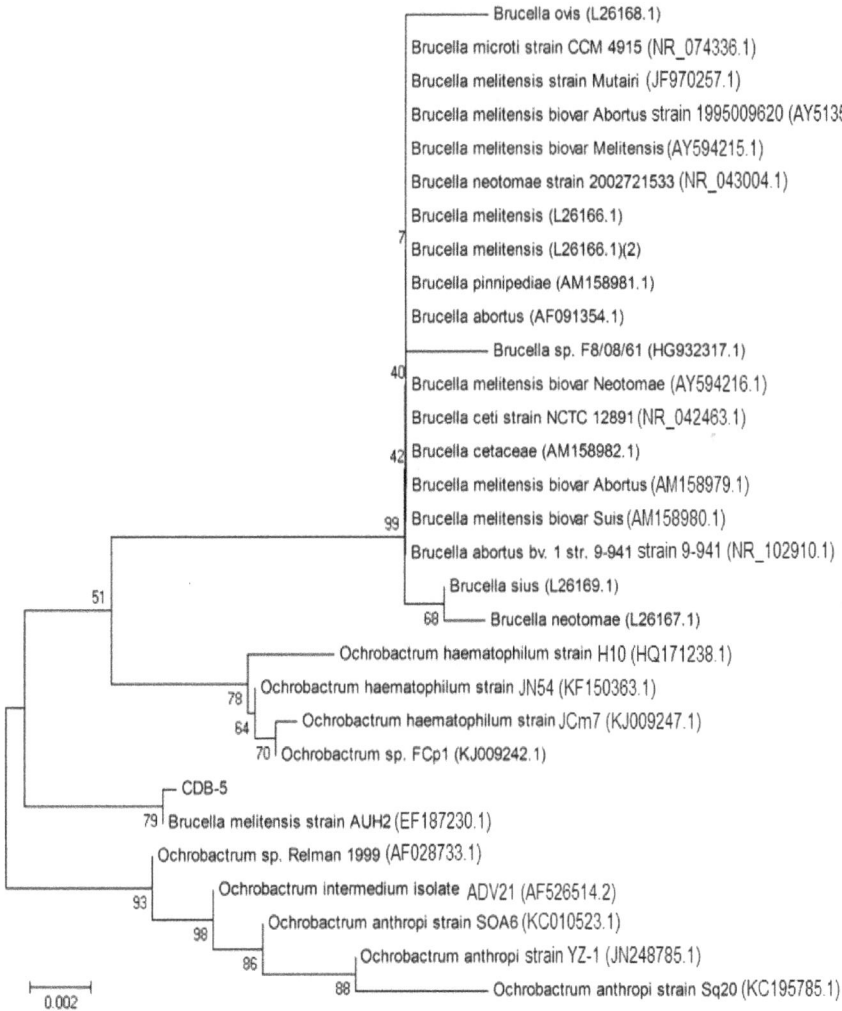

Fig. 35: Phylogenetic tree of *Brucella* sp. (CDB-5) isolated from mangrove soil of Mahanadi river delta.

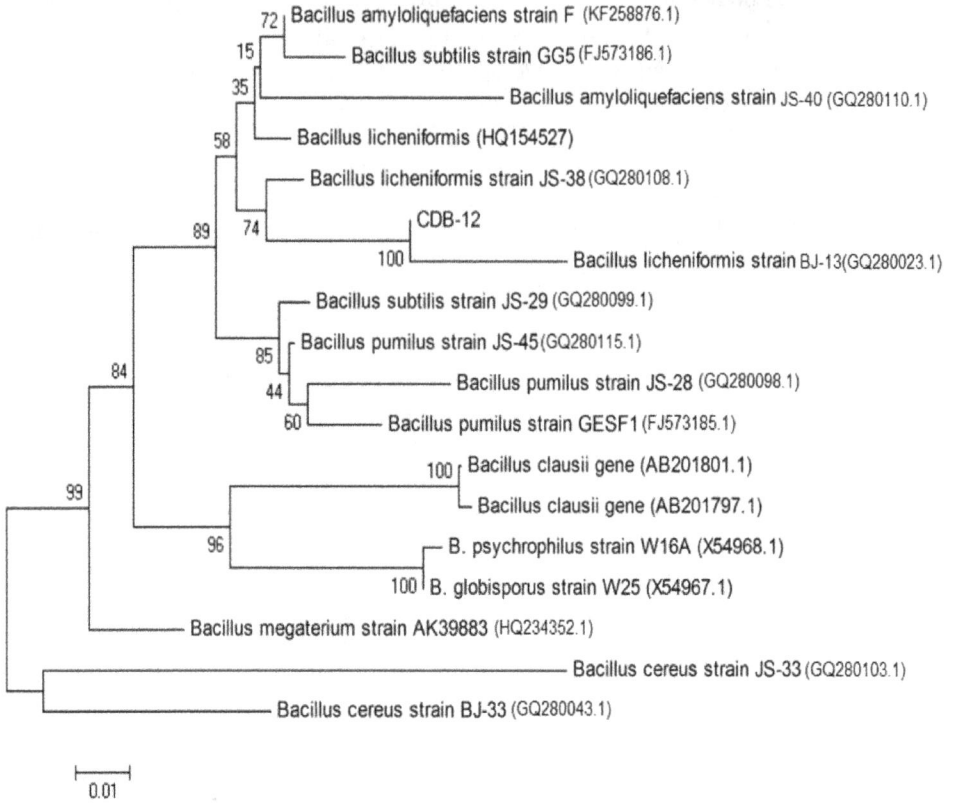

Fig. 36: Phylogenetic tree of *Bacillus* spp. (CDB-12) isolated from mangrove soil of Mahanadi river delta.

Conclusion

The 29 bacterial species identified from four selected bacterial groups along with isolation and identification of 27 fungal and 11 algal species revealed that the mangrove of Bhitarkanika is endowed with rich microbial diversity. Among the bacterial genera, *Bacillus*, *Pseudomonas* and *Desulfotomaculum* were dominant in the mangrove soils. The bacteria from mangrove soil possess unique biotechnological potentials like N_2 fixation, cellulose degradation, phosphate solubilization as well as sulfur oxidation. Information of different groups of microorganisms from the saline habitat of Bhitarkanika will not only help to assess the microbial diversity, but also will provide information to explore the important microorganisms for various biotechnological applications. Like Bhitarkanika, mangroves of Mahanadi delta is also enrich in microbial population not in number but in type. The study made by the authors on microbial diversity from mangroves of Mohanadi delta revealed occurrence of number of algae, fungi and bacteria. Altogether, 12 algal, 20 fungal and 35 bacterial isolates were isolated from mangroves of Mahanadi delta. Among the bacterial strains *Bacillus*, *Micrococcus* and *Pseudomonas* sp. were found dominat in the mangrove soil of Mahanadi delta.

4

BIOTECHNOLOGICAL POTENTIAL OF MANGROVE MICROORGANISMS

4.1 Biotechnological Potential of Mangrove Microorganisms

Microbial diversity is the key to human survival and economic wellbeing and provides a huge reservoir of resources, which we can utilize for our benefit. Mangrove microorganisms proven to be an important sources of food, feed, medicine, enzymes and antimicrobial substances (Lin *et al.*, 2001; Maria *et al.*, 2005). Both halotolerant and halophilic bacteria and other microbes from mangrove ecosystem has large number of industrial applications in terms of their unique enzymes (Sabu, 2003), capable to produce biosurfactants (Yakimov *et al.*, 1999), bioplastics (Steinbuchel *et al.*, 1998), compatible solutes (Margesin and Schinner, 2001), natural bioproduct and other commercially important products. Mangrove fungi is the principal commercial sources of xylanolytic enzymes which have many industrial uses, such as in paper manufacturing, animal feed, bread making, juice preparation, wine industries and xylitol production (Polizeli *et al.*, 2005). Marine algae are the only sources for industrially important phycocolloids like agar, carrageenan and alginate (Shanmugan and Mody, 2000). They are also reported to have blood anticoagulant, anti-tumour, anti-mutagenic, anti-complementary, immunomodulating, hypoglycemic, antiviral, hypolipidemic and anti-inflammatory activities. Actinomycetes isolated from mangrove habitats are potentially rich source for the discovery of anti-infection and anti-tumor compounds and of agents for treating neurodegenerative diseases and diabetes (Hong *et al.*, 2009). Besides the above approach, mangrove microorganisms have wide applications in agriculture, industry as well as production of various secondary metabolites for human use which are described below.

4.1.1. Agricultural application

Microorganisms of agricultural importance represent key ecological strategy for integrated management practices like nutrient management, disease and pest management in order to reduce the use of chemicals in agriculture as well as to improve cultivar performance. It has been already reported that the mangrove microorganisms are beneficial for agriculture (Kathiresan and Selvam, 2006). The rhizosphere soil of mangrove plants harbors a large number of beneficial bacteria with a large number of agricultural applications (Kathiresan and Selvam, 2006). These strains have the ability to (1) fix nitrogen, (2) solubilize phosphate, (3) produce ammonia, and (4) produce the plant growth hormone indole acectic acid (IAA).

Bacterial strains identified from rhizosphere soil of mangroves such as, *Azotobacter vinelandii* and *Bacillus megaterium* have shown their ability to enhance mangrove seedlings (Kathiresan and Selvam, 2006). Soil bacteria present in root regions are known to enhance plant growth. The N_2- fixing bacteria isolated from saline soil could be good candidates for use to improve the fertility of reclaimed arid and saline soils (Zahran *et al.* 1995). The inoculation of plants with plant-growth-promoting bacteria is a common tool in agriculture to enhance crop yields (Bashan and Holguin, 1997). Mangrove soils also contain phosphate solubilizing microrganisms (PSB). These microrganisms have great scope for their application as biofertilizer for improving plant growth and production in saline soils.

4.1.2. Industrial applications

4.1.2.1. *Enzymes*

Mangrove ecosystem provides the unique habitat for the colonization of fungi and bacteria. Interestingly, the enzymes derived from mangrove associated microrganisms have enormous economic value in industries of agriculture, pulp, paper, medicine and sewage treatment etc. Many researchers have shown the importance of the mangrove derived fungi on biotechnological applications and industrial uses. Especially the mangrove associated fungal derived enzymes are applied in paper manufacturing, juice and wine industries, bread making, xylitol production, bioremediation of dye and metal degredation/removal (Polizeli *et al.*, 2005; kathiresean *et al.*, 2011).

Mangrove microorganisms have a diverse range of enzymatic activity and are capable of catalyzing various biochemical reactions with novel enzymes. Especially, halophilic microorganisms possess many hydrolytic enzymes (amylases, nucleases, phosphatases, and proteases) and are capable of functioning under conditions that lead to precipitation of denaturation of most proteins (Ventosa and Nieto, 1995). Sea water, which is saline in nature, could provide microbial products, in particular the enzymes that could be safer having no or less toxicity of side effects when used for therapeutic applications to humans (Sabu, 2003). The hydrolases enzymes produced by halophilic bacteria are currently of commercial interest (Ventosa and Nieto, 1995). Halophilic bacteria (*Halococcus*) produce L-asperginase was also reported from mangrove environment (Sudha, 1981). From Brazil mangrove sediment *Vibrionales* appeared to be the predominant enzyme-producing group within the community when compared with other groups (*Actinomycetales* and *Bacillales*), mainly for the production of amylase and protease (Dias *et al.*, 2009). The order *Vibrionales* revealed to be metabolically versatile with a high production of enzymes. An isolate of *Vibrio fluvialis* from mangrove sediments was used to produce an alkaline extracellular protease with high efficiency for use in industrial detergents (Venugopal and Saramma, 2006). Mishra *et al.* (2010) evaluated and reported the activity of some stress enzymes such as catalase, peroxidase, oxidase, polyphenol oxidase and ascorbic acid oxidase from six gram negative bacteria isolated from mangroves of Bhitarkanika, Odisha, India. Bacteria isolated from mangrove sediment of Brazil were found to produce diverse extracellular enzymes such as amylase, protease, esterase and lipases (Armando *et al.*, 2009). Joseph and Paul Raj (2007) reported five phytase producing *Bacillus* strains from mangrove ecosystem of Kochin, Kerala, India. Three bacterial and one fungal strain producing tanase have been isolated from the mangrove forest of North malbar, Kerla, India. Wu (1993) identified 15 genera (42 strains) of fungi from mangroves in the Tansui Estuary near Taipei, Taiwan, and found that most of the *Ascomycetes* were able to secrete a wide range of enzymes potentially capable of decomposing mangrove litter. Raghukumar *et al.* (2004) reported that a mangrove fungus, *Aspergillus niger*, can produce thermostable, cellulose-free alkaline xylanase that showed activity in biobleaching of paper pulp and the crude enzyme of its with high xylanase activity and could bring about bleaching of sugarcane bagasse pulp by a 60-minutes treatment at 55 °C. A marine hypersaline-tolerant white-rot fungus, *Phlebia* sp. MG-60, screened from mangrove stands (Xin *et al.*, 2002), has shown excellent lignin degradation ability. It can degrade more than 50% of lignin incubated with whole sugarcane bagasse and the whole sugarcane bagasse might be used to produce animal feed after fermentation (Xin *et al.*, 2003).

4.1.2.2. Biosurfactants

Biosurfactants have advantages over their chemical counterparts because they are bio-degradable, have low toxicity, are effective at extreme temperatures or pH values and show better environmental compatibility (Mulligan, 2009). Recently the *Leucobacter komagatae* 183 strain, isolated from mangrove sediment in Trang, southern Thailand, was evaluated as a potential biosurfactant producer (Saimmai *et al.*, 2011).

4.1.3. Pharmaceutical applications

The needs for the diversity and development of new classes of antimicrobial compounds are increasing, due to trends in antibiotic resistance among different strains of bacteria, fungi and other microorganism. A significant number of reports focused on antimicrobial metabolites isolated from mangrove saprophytic fungi. A fungal strain *Preussia aurantiaca* isolated from mangrove forest was found to synthesize two new despidones (Auranticins A and B) which display antimicrobial activity (Poch and Gloer, 1991). Aigialomycins A-E, new 14-membered resorcylic macrolides, was isolated together with a known hypothemycin from the mangrove fungus, *Aigialus parvus* BCC 5311 (Isaka *et al.*, 2002). Enniating G, a new compound with a structure of cyclohexapeptide, was also isolated from the culture broth of mangrove fungus *Fusarium* sp. (Lin *et al.*, 2002a; Lin & Zhou 2003; You *et al.*, 2006), which displayed antitumor, antibiotic, insecticidal and phytotoxic activity. An ascomycete, *Verruculina enalia*, is a common tropical species found on mangrove wood worldwide, reported to produce two new phenolic compounds, enalin A and B, with hydroxylmethyl furfural and three cycloidpeptides from its fermentation broth. Enalin A is a coumaranone, a type of compound distributed widely from microorganisms to higher plants and having antimicrobial, antifungal, phytotoxic and antidiabetic activities (Lin *et al.*, 2002b). Among the mangrove fungi, more and more mangrove endophytes now have been researched and more and more antimicrobial metabolites have been isolated. One of them is Cytosporone B which shows broad activities against fungi. In addition to fungi Wiwin (2010) reported an actinomycete *Steptomyces* sp. from mangrove soil in the eastern coast of Surabaya, Indonesia, capable of producing a series of antibiotics that strongly inhibit the growth of Gram positive and Gram negative bacteria. Santhi and Jebakumar (2011) reported some *Streptomyces* sp. from mangrove sediment of Manakudi estuary, India, exhibits potent antimicrobial effects against methicillin-resistant *Staphylococcus aureus* (clinical isolate) and methicillin-susceptible *S. aureus* and *Salmonella typhi*.

4.1.3.1. *Bioactive compounds*

Mangrove ecosystems have been considered a "hot-spot" for newer and better drugs naturally produced by the microorganisms living in this environment. It is encouraging that bioactive compounds have been obtained from mangrove plants fungi, bacteria including actinomycetes (Cheng *et al.*, 2009). Actinomycetes isolated from mangrove habitats are a potentially rich source for the discovery of anti-infection and anti-tumor compounds, and of agents for treating neurodegenerative diseases and diabetes (Hong *et al.*, 2009). In the mangroves situated around the coast of China and surrounding islands, Hong *et al.* (2009) compiled more than 2,000 actinomycetes with the potential to synthesize biologically active secondary metabolites that conferred anti-tumor, anti-cancer and anti-infection properties. Using molecular and morphological/biochemical identification techniques and screening tests, approximately 20% showed activity against the growth of Human Colon Tumor 116 cells, whereas only 3% inhibited the protein PTP1B associated with diabetes. Lin *et al.* (2001) also reported three new 2-pyranon derivatives from the mangrove actinomycetes *Nocardiopsis* sp. A00203. Sivakumar *et al.* (2005) reported a Streptomyces *albidoflavous* from the Pichavaram mangrove which showed antitumor properties. Five unique metabolites, xyloketals A (1), B (2), C (3), D (4), and E (5) were isolated from mangrove fungus *Xylaria* sp. (no. 2508), obtained from the South China Sea. Tao *et al.* (2010) reported Isoflavone and Prostaglandin analog compound from mangrove fungi isolated from South China Sea which appeared to be promising for treating cancer patients with multidrug resistance. Huang *et al.* (2011) reported eight secondary metabolites including three new azaphilones (chermesinones A–C, 1–3), three new *p*-terphenyls (6'-O-desmethylterphenyllin, 4; 3-hydroxy-6'-O-desmethylterphenyllin, 5; 3''-deoxy-6'-O-desmethylcandidusin B, 7), and two known *p*-terphenyls (6, 8), from the culture of the mangrove endophytic fungus *Penicillium chermesinum* (ZH4-E2). Eight new indole triterpenes named shearinines D-K, along with shearinine A, paspalitrem A, and paspaline, have been isolated from the mangrove endophytic fungus *Penicillium* sp. Shearinines D, E, and (with reduced potency) G exhibit significant *in vitro* blocking activity on large-conductance calcium-activated potassium channels (Xu *et al.*, 2007). β-carboline, adenosine and 8-hydroxyl-3,5-dimethyl -isochroman-1-one, were isolated from mangrove fungus K32. The interaction of β-carboline with calf thymus DNA was investigated by UV-vis and fluorescence spectra, resulting in the occurrence of binding reaction, which was proposed to be one possible mechanism of the antitumor activity of β-carboline (Song *et al.*, 2004).

4.1.4. Environmental applications

Bioremediation strategies can be improved by a greater knowledge of microbiology and new molecular technologies that can support this development. Genomics, for instance, provides complete genome sequence data for several microorganisms that are significant for bioremediation, such as *Pseudomonas, Shewanella, Sphingomonas, Arthrobacter* etc. (Desai *et al.*, 2010). In addition to processing nutrients, mangrove bacteria may also help in processing the industrial wastes. Iron reducing bacteria were common in mangrove habitats in some mining areas (Panchanadikar, 1993). Eighteen bacterial isolates that metabolise waste drilling fluid were collected from a mangrove swamp in Nigeria (Benka-coker and Olumagin, 1995). The presence and activity of the oil degrading microorganisms in the mangrove sediments not only plays a key roles of bioremediation of oil in mangroves but also considered to be a new prospects for the use of molecular tools to monitor the bioremediation process. Microorganisms are directly involved in biogeochemical cycles as key drivers of the degradation of many carbon sources, including petroleum hydrocarbons. Various bacterial groups present in mangrove sediment are already known for their capacity to degrade hydrocarbons, such as *Pseudomonas, Marinobacter, Alcanivorax, Microbulbifer, Sphingomonas, Micrococcus, Cellulomonas, Dietzia* and *Gordonia* (Brito *et al.*, 2006). Bacteria were isolated in the sediment of a mangrove located in Hong Kong, China, which demonstrated a great capacity for Poly aromatic hydrocarbon (PAH) degradation *in vitro* and could be used to degrade PAH in contaminated sediment (Ramsay *et al.*, 2000). Similarly, Yu *et al.* (2005) investigated the biodegradability of PAHs, fluorine (Fl), phenanthrene (Phe) and pyrene by a bacterial consortium enriched with mangrove sediment. The consortium was formed by three bacterial strains: *Rhodococcus* sp., *Acinetobacter* sp. and *Pseudomonas* sp. Ramsay *et al.* (2000) reported a large number of and a wide diversity of PAH-degrading microorganisms in Australian mangrove sediments. Ke *et al.*, (2003), in a study that used contaminated mangrove microcosms, demonstrated the removal of 90% of pyrane in 6 months. Some of the bacterial species such as *Streptococcus, Staphylococcus, Micrococcus, Moraxella* and *Pseudomonas* and fungal species such as *Aspergillus gloocus* and *A.niger* were also reported to degrade polythene and plastic bags from the mangrove environment (Kathiresan, 2003). D'Souza *et al.* (2006) reported that a mangrove white-rot basidiomycetous fungus, able to produce laccase to decolorize colored effluents and synthetic dyes. The efficiency of this fungus in decolorization of various effluents with laccase that is active at pH 3.0-6.0 and at 60 °C in the presence of seawater has great potential in bioremediation of industrial effluent. Enhanced laccase production in the presence of industrial effluents in this fungus is an added advantage during bioremediation of effluents. Two moderately halotolerant *Bacillus megaterium* species isolated from mangroves of Bhitarkanika, Odisha showed potentiality for reduction of toxic selenite to non-toxic elemental selenium (Mishra *et al.*, 2011). Similarly a strain of *Vigribacillus* sp. isolated from mangrove soil of Bhitarkanika showed potential chromate reduction ability (Mishra *et al.*, 2012).

4.2. Biotechnological Potential of Bacteria Isolated from Mangroves of Odisha Coast

4.2.1. Phosphate solubilizing bacteria (PSB) from Mahanadi delta

Phosphate solubilizing bacteria as potential suppliers of soluble phosphorus should confer a great advantage for plants through solubilization and mineralization (Rodriguez & Fraga, 1999). Mangrove soils, due to their normally high organic content may possess a high proportion of organic-bound P, up to 75–80% of the total extractable P (Alongi *et al.*, 1992). The anoxic conditions of the sediments beneath the aerobic zone would generally tend to favor dissolution of nonsoluble phosphate through sulfide production. In an arid mangrove ecosystem in Mexico, nine strains of phosphate-solubilizing bacteria were isolated from black mangrove (*A. germinans*) roots; *Bacillus amyloliquefaciens, B. atrophaeus, Paenibacillus macerans, Xanthobacter agilis, Vibrio proteolyticus, Enterobacter aerogenes, E. taylorae, E. asburiae, Kluyvera cryocrescens*, and three strains from white mangrove (*Languncularia racemosa*) roots; *B. licheniformis, Chryseomonas luteola*, and *Pseudomonas stutzeri* (Vazquez *et al.*, 2000).

Solubilization of mineral phosphate by phosphate solubilizing bacteria is generally associated with the release of low molecular weight organic acids (Goldstein, 1995). Some of the organic acids might act as chelators displacing metals from phosphate complexes. Solubilization of phosphate-rich compounds is also carried out by the action of a phosphatase enzyme called phosphatase. In recent years, different screening programs have been performed in saline habitats for isolation of phosphate solubilising microorganisms. Some phosphate-solubilizing bacteria have been identified and characterized from saline mangrove soils of the Mahanadi river delta, Odisha and attempt has been made to purify and characterize the alkaline and acid phosphatase enzyme produced by the bacterial isolates which may have potential biotechnological application. In addition, the efficiency of the isolated and identified bacterium was also evaluated in relation to its plant growth promotion.

4.2.1.1. Isolation and screening of PSB

Phosphate solubilising bacteria was isolated from six different location of mangrove ecosystem. For this purpose soil was homogenized in sterile Milli Q water containing 0.85% NaCl (w/v), serially diluted (10^{-4}) and spreaded on pikovskaya's agar medium plates. Then the petriplates were incubated at 30 ^{0}C for 24-48 h. Colonies were selected from the plates on the basis of the appearance of a clear halozone (**Fig.33**). Forty eight morphologically distinct bacterial isolates forming halozones in Pikovskaya-agar medium were selected as phosphate solubilising bacteria (PSB1-48). Pure culture of these bacteria were prepared and maintained in the laboratory for further studies. These bacteria were subjected to further qualitative and quantitative screening for evaluation of their phosphate solubilisation abilities. For screening, National Botanical Research Institute Pune (NBRIP) liquid medium (Nautiyal, 1999); (gL^{-1}) ($MgCl_2.6H_2O$ (5), $MgSO_4.H_2O$ (0.25), KCl (0.2), $(NH_4)_2SO_4$ (0.1), $Ca_3(PO4)_2$ (5) glucose (10 g) was inoculated with a 1% (v/v) inoculum from 48 different strains of pre-grown cultures in the same medium. In all the experiments tricalcium phosphate $Ca_3(PO_4)_2$ was used as the sole sources of Phosphourus (P).

Flasks containing 100 ml inoculated medium were incubated at 30 °C on a shaker maintained at 100 rpm. From each inoculated flask 2 ml of samples were taken aseptically at regular intervals for further studies.

4.2.1. 2. Qualitative screening of phosphate solubilizing abilities

Qualitative screening of phosphate solubilising ability of selected fourty eight bacterial isolates (PSB1-48) previously forming halozones in Pikovskaya-agar medium were carried out in terms of changes of colour intensity in NBRIP-BPB broth. The results of quantitative study are given in **Table-22 and Fig. 37.** Phosphate solubluilizing abilities of the bacterial isolates measued in terms of changes of colour intensities of bromophenol blue in NBRIP-BPB broth showed great variations among the strains. The value of spectrophotometric reading ranged from 0.025 (PSB-47) to 1.188 (PSB-26) O.D. at 600 nm. The maximum color change as obtained with the isolate, PSB-26 was 1.188 and PSB-37 with vale 1.145. Further, confirmation of phosphate solubilising abilities of these 48 bacteria was also made by observing the formation of holozones by the bacteria in NBRIP-agar medium again by calculating their phosphate solubilising efficiency (Zone size / Colony size x 100). Their phosphate solubilising efficiency were found to vary from 108.0 -175.0. Maximum phosphate solubilising efficiency was shown by the isolate PSB-37 (175), followed by the isolate PSB-26 (170) while the least value was observed with the isolate, PSB-5 (108).

Table 22: Comparative evaluation of tricalcium phosphate solubilisation

Isolates No.	Change in colour intensity	Colony size (cm)	Zone Size (cm)	P solubilising Efficiency (Zone size) / Colony size x 100
PSB-1	0.186 ± 0.002	0.6	0.7	116.66
PSB-2	0.265 ± 0.005	0.8	0.9	112.5
PSB-3	0.362 ± 0.003	1.0	1.1	110.00
PSB-4	0.379 ± 0.006	0.3	0.4	133.33
PSB-5	0.253 ± 0.001	1.2	1.3	108.33
PSB-6	0.088 ± 0.004	1.0	1.1	110.00
PSB-7	0.175 ± 0.006	0.8	0.9	112.50
PSB-8	0.263 ± 0.003	0.7	0.8	114.28
PSB-9	0.365 ± 0.003	0.5	0.6	120.00
PSB-10	0.372 ± 0.002	0.5	0.6	120.00
PSB-11	0.123 ± 0.005	0.3	0.4	133.33
PSB-12	0.655 ± 0.003	1.0	1.3	130.00
PSB-13	0.453 ± 0.006	0.8	0.9	112.50
PSB-14	0.373 ± 0.001	0.6	0.7	116.66

Isolates No.	Change in colour intensity	Colony size (cm)	Zone Size (cm)	P solubilising Efficiency (Zone size) / Colony size x 100)
PSB-15	0.662 ± 0.004	1.3	1.6	123.00
PSB-16	0.841 ± 0.006	1.1	.1.5	136.36
PSB-17	0.437 ± 0.003	0.6	0.7	116.66
PSB-18	0.692 ± 0.003	0.6	1.0	166.66
PSB-19	0.087 ± 0.002	1.0	1.2	120.00
PSB-20	0.343 ± 0.005	0.7	0.9	128.57
PSB-21	0.712 ± 0.003	0.4	0.5	125.00
PSB-22	0.178 ± 0.006	1.1	1.2	109.09
PSB-23	0.369 ± 0.001	0.8	0.9	112.50
PSB-24	0.813 ± 0.004	0.5	0.6	120.00
PSB-25	0.178 ± 0.006	1.1	1.3	118.18
PSB-26	1.188 ± 0.003	1.0	1.7	170.00
PSB-27	0.833 ± 0.003	0.3	0.4	133.11
PSB-28	0.935 ± 0.002	0.3	0.4	133.00
PSB-29	1.123 ± 0.005	0.8	1.2	150.00
PSB-30	0.259 ± 0.003	0.5	0.6	120.00
PSB-31	0.276 ± 0.006	0.6	0.7	116.66
PSB-32	0.372 ± 0.001	0.3	0.4	133.33
PSB-33	0.140 ± 0.004	0.9	1.0	111.11
PSB-34	0.518 ± 0.006	0.7	0.9	128.57
PSB-35	0.288 ± 0.003	0.5	0.6	120.00
PSB-36	0.399 ± 0.003	0.4	0.5	125.00
PSB-37	1.145 ± 0.002	0.8	1.4	175.00
PSB-38	0.333 ± 0.005	0.5	0.6	120.00
PSB-39	0.161 ± 0.003	0.3	0.4	133.33
PSB-40	0.789 ± 0.006	0.7	0.8	114.28
PSB-41	0.835 ± 0.001	0.5	0.6	120.00
PSB-42	0.393 ± 0.004	0.6	0.7	116.66
PSB-43	0.225 ± 0.006	0.7	0.8	114.28

Isolates No.	Change in colour intensity	Colony size (cm)	Zone Size (cm)	P solubilising Efficiency (Zone size) / Colony size x 100)
PSB-44	0.893 ± 0.003	0.4	0.5	125.00
PSB-45	0.079 ± 0.003	0.8	0.9	112.5
PSB-46	0.286 ± 0.002	1.0	1.2	120.00
PSB-47	0.025 ± 0.005	1.2	1.3	108.33
PSB-48	1.085±0.003	0.4	0.5	125.0

4.2.1.3. Selection of efficient phosphate solubilizing bacterial isolates

Taking into consideration the scoring of phosphate solubilizing efficiency and changes in color intensities of both in NBRIP agar and NBRIP-BPB broth, all the 48 bacterial isolates were ranked (**Tabe-23**).

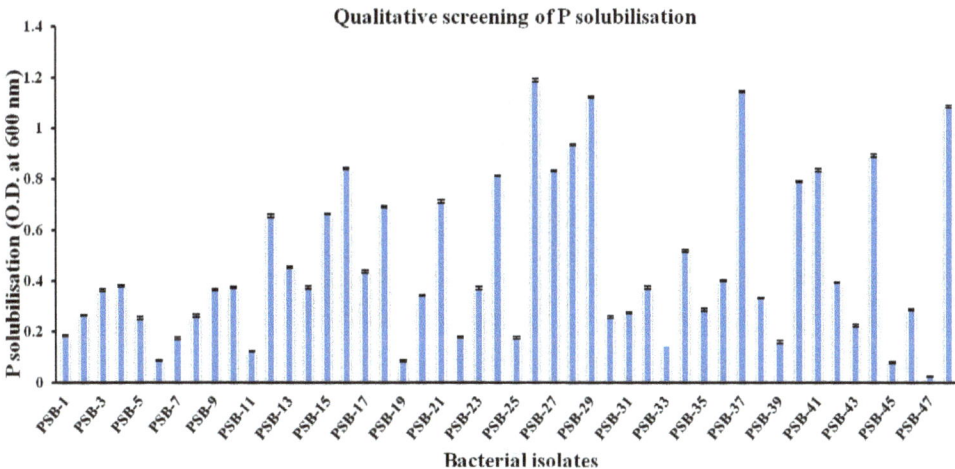

Fig. 37: Comparative P solubilisation ability in terms of change in color intensitiy of NBRIP- BPB broth medium by different P solubilising bacterial isolate

Table 23: Scoring of P solubilizing ability of phosphate solubilizing bacteria isolated from mangrove soil of Mahanadi delta, Odisha.

Isolates	P solubilising Efficiency	Score towards (P solubilising Efficiency) (a)	Change in colour intensity	Score towards Change in colour intensity after 72 hh (b)	Ranking in P solubilising ability in terms of total score (a+b)
PSB-1	116.66	14	0.186 ± 0.002	34	48
PSB-2	112.5	16	0.265 ± 0.005	30	46
PSB-3	111.11	17	0.362 ± 0.003	24	41
PSB-4	133.33	6	0.379 ± 0.006	41	47

Isolates	P solubilising Efficiency	Score towards (P solubilising Efficiency) (a)	Change in colour intensity	Score towards Change in colour intensity after 72 hh (b)	Ranking in P solubilising ability in terms of total score (a+b)
PSB-5	108.33	19	0.253 ± 0.001	33	52
PSB-6	110	18	0.088 ± 0.004	40	58
PSB-7	112.5	16	0.175 ± 0.006	36	52
PSB-8	114.28	15	0.263 ± 0.003	31	46
PSB-9	120	12	0.365 ± 0.003	23	35
PSB-10	120	12	0.372 ± 0.002	21	33
PSB-11	133.33	7	0.123 ± 0.005	39	46
PSB-12	130	8	0.655 ± 0.003	14	22
PSB-13	112.5	16	0.453 ± 0.006	16	32
PSB-14	116.66	14	0.373 ± 0.001	20	34
PSB-15	123	11	0.662 ± 0.004	13	24
PSB-16	136.36	5	0.841 ± 0.006	7	12
PSB-17	118	13	0.437 ± 0.003	17	30
PSB-18	166.66	3	0.692 ± 0.003	12	15
PSB-19	120	12	0.087 ± 0.002	41	53
PSB-20	128.57	9	0.343 ± 0.005	25	34
PSB-21	125	10	0.712 ± 0.003	11	21
PSB-22	110	18	0.178 ± 0.006	35	53
PSB-23	112.5	16	0.369 ± 0.001	22	38
PSB-24	120	12	0.813 ± 0.004	9	21
PSB-25	118	13	0.178 ± 0.006	35	48
PSB-26	170	2	1.188 ± 0.003	1	3
PSB-27	133.33	7	0.833 ± 0.003	8	15
PSB-28	133.33	7	0.935 ± 0.002	5	12
PSB-29	150	4	1.123 ± 0.005	3	7
PSB-30	120	12	0.259 ± 0.003	32	44
PSB-31	116.66	14	0.276 ± 0.006	29	43
PSB-32	133.33	7	0.372 ± 0.001	21	28
PSB-33	111.11	17	0.140 ± 0.004	38	55
PSB-34	128.57	9	0.518 ± 0.006	15	24
PSB-35	120	12	0.288 ± 0.003	27	39

Isolates	P solubilising Efficiency	Score towards (P solubilising Efficiency) (a)	Change in colour intensity	Score towards Change in colour intensity after 72 hh (b)	Ranking in P solubilising ability in terms of total score (a+b)
PSB-36	125	10	0.399 ± 0.003	18	28
PSB-37	175	1	1.145 ± 0.002	2	3
PSB-38	120	12	0.333 ± 0.005	26	38
PSB-39	133.33	7	0.161 ± 0.003	37	44
PSB-40	114.28	15	0.789 ± 0.006	10	25
PSB-41	120	12	0.835 ± 0.001	8	20
PSB-42	116.66	14	0.393 ± 0.004	19	31
PSB-43	116.66	14	0.225 ± 0.006	34	48
PSB-44	125	10	0.893 ± 0.003	6	16
PSB-45	114.28	15	0.079 ± 0.003	42	57
PSB-46	120	12	0.286 ± 0.002	28	40
PSB-47	108.33	19	0.025 ± 0.005	43	62
PSB-48	125	10	1.085 ± 0.003	4	14

The bacterial isolates were scored 48 to 1 in order of their decreased value towards P solubilising efficiencies and changes in color intensities. Ranking was done based on the combined score obtained by a particular bacterium towards P solubilising efficiencies and changes of color intensities. Based on their ranking, fourteen out of the 48 isolates (PSB-12, PSB-15, PSB-16, PSB-18, PSB-21, PSB-26, PSB-27, PSB-28, PSB-29, PSB-34, PSB-37, PSB-41, PSB-44 and PSB-48) were selected as efficient (having rank 1-14) phosphate solubilizing bacteria (**Table 23**) and were selected for quantitative phosphate estimation.

4.2.1.4. Quantitative estimation of available/soluble phosphate

Further quantitative evaluation of P solubilising abilities of selected fourteen bacterial isolates was carried out spectrophotometrically for a period of 264 h by inoculating in NBRIP-broth medium containing tricalcium phosphate. The quantitative estimation of phosphate, solubilized by each isolates was determined by the method of Murphy and Riely (1962). Erlenmeyer flasks containing 100 ml medium were inoculated with bacteria. Uninoculated medium served as the control. The flasks were incubated in a shaker at 30 °C upto 264 h at a shaking speed of 100 rpm. The pH of the culture medium was measured at specific intervals of incubation period. The culture was collected at every 24 h interval and centrifuged at 10,000 rpm for 10 minutes. The supernatant was separated from the bacterial cells by successive filtration through Whatman no.1 filter paper followed by 0.22 µm millipore membrane and used to estimate the phosphate release in triplicate spectrophotometrically at 880 nm. The amount of phosphate released was compared with the standard curve taking specific amount of potassium dihydrogen phosphate as standard. For quantitative estimation of available/soluble phosphate, 50.0 ml of sample was pipeted into a clean, dry test tube or 125 ml Erlenmeyer flask and 0.05 ml (1 drop) of phenolphthalein indicator was added to this. $5N$ H_2SO_4 solution was added dropwise to discharge the red color (if red color developed). To each sample, 8.0 ml combined reagent was then added and mixed thoroughly. After at least 10 mins but not more than 30 min (to develope the colour), the absorbance of each sample was taken at 880 nm, using combined reagent as the reference solution. In this method orthophosphate reacted with molybdate to form phosphomolybdic acid which is reduced by ascorbic acid to form a blue complex. Amount of available phosphate was estimated from the standard curve of available phosphate.

Results of the quantitative P evaluation are presented in Table 24. Simultaneously, with the quantitative evaluation of P solubilisation the reduction of pH of the medium (from initial pH of 7.0) were also observed up to 264 h of incubation and the data are presented in Table 25. The soluble P concentrations in the medium were found in the range of 8.21 and 48.70 µg/ml with variation among different isolates at different incubation periods. Tricalcium phosphate solubilisation by the bacteria increased with increase of incubation time up to a certain period thereafter it declined. The maximum P solubilising abilities by different bacteria were observed at different incubation periods. The graphical presentation of maximum P solubilising abilities and maximum changes of pH of the medium are given in Fig. 38 and 39 respectively. The maximum P solubilisation was observed by the isolate PSB-26 (48.70 µg/ml) at 144 h of incubation followed by PSB-37 with 44.84 µg/ml after 72 hour of incubation (Fig. 40 and 41). In addition to phosphate solubilising ability, the maximum dropping of pH was also observed by the isolate PSB-37 (3.15) followed by PSB-26 (3.23) after 72 hour of incubation. In the control no soluble-P was detected as well no drop in pH was observed.

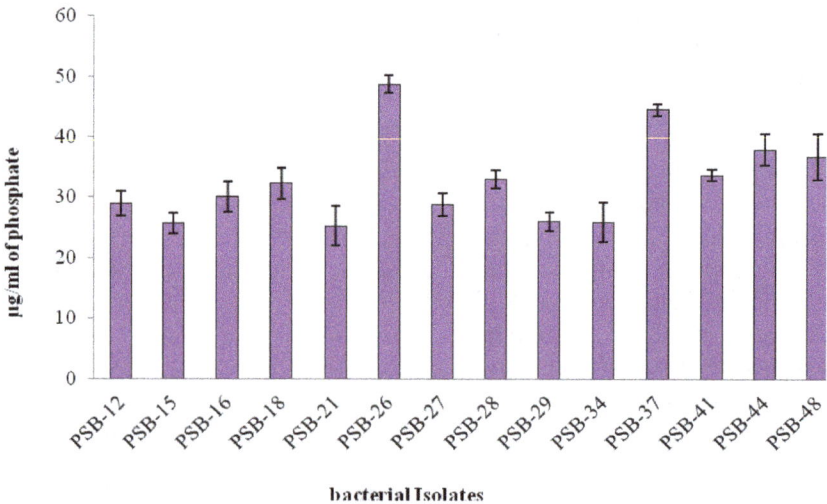

Fig. 38: Phosphate solubilising efficiency of fourteen bacterial isolates from mangrove soil of Mahanadi delta

Table 24: Tricalcium Phosphate solubilized (µg/ml) by selected phosphate solubilising bacterial isolate during 264 h of incubation in NBRIP-broth medium

Isolates	6 h	12 h	24 h	48 h	72 h	96 h	120 h	144 h	168 h	192 h	216 h	240 h	264 h
PSB-12	12.80 ±2.17	17.71 ±2.11	22.82 ±2.21	23.08 ±2.07	28.92 ±2.08	29.18 ±1.88	34.71 ±1.76	35.50 ±1.59	38.69 ±1.23	37.32 ±0.97	35.88 ±1.6	34.75 ±1.54	34.8 ±0.85
PSB-15	10.38 ±1.42	11.0 ±1.65	17.09 ±1.64	21.88 ±1.43	25.68 ±1.76	25.66 ±1.87	35.59 ±2.53	36.42 ±1.21	36.69 ±2.75	32.91 ±1.2	32.56 ±1.4	31.50 ±1.65	31.23 ±1.75
PSB-16	13.62 ±2.71	14.29 ±2.43	21.65 ±2.76	31.87 ±2.65	31.97 ±2.54	38.0 ±2.32	37.85 ±2.21	37.76 ±1.22	35.17 ±1.32	35.26 ±1.81	31.92 ±1.43	30.51 ±2.44	30.42 ±2.41
PSB-18	10.64 ±1.31	11.11 ±1.76	21.83 ±2.43	32.00 ±3.12	32.29 ±2.54	39.79 ±3.59	33.13 ±3.29	33.96 ±2.28	30.49 ±1.93	30.44 ±1.65	30.07 ±1.20	28.18 ±2.49	27.15 ±1.59
PSB-21	19.17 ±1.58	19.28 ±1.31	20.25 ±2.63	22.56 ±1.75	25.27 ±3.31	26.99 ±2.83	28.25 ±1.64	29.53 ±2.33	33.14 ±1.88	32.10 ±3.01	32.56 ±3.54	31.06 ±2.52	30.00 ±2.43
PSB-26	13.51 ±2.66	24.29 ±3.54	37.21 ±1.64	40.45 ±2.77	45.10 ±1.43	45.85 ±1.21	46.52 ±1.29	48.70 ±1.28	44.57 ±1.41	43.06 ±1.44	40.75 ±1.36	36.53 ±1.26	36.45 ±1.16
PSB-27	26.16 ±2.18	26.33 ±1.18	27.44 ±2.17	27.88 ±2.51	28.75 ±1.83	28.86 ±1.52	29.94 ±1.15	30.33 ±3.32	31.65 ±1.42	31.66 ±2.88	33.94 ±1.33	33.74 ±2.61	32.29 ±1.52
PSB-28	11.88 ±1.65	22.40 ±1.54	32.62 ±1.64	32.99 ±1.87	33.06 ±1.53	33.26 ±1.43	34.68 ±1.66	35.48 ±1.98	37.35 ±3.88	35.08 ±2.43	20.49 ±1.56	20.66 ±2.6	20.44 ±2.33
PSB-29	11.04 ±1.53	11.22 ±1.42	17.81 ±2.77	18.12 ±1.43	26.03 ±1.54	27.11 ±1.21	27.47 ±1.61	28.07 ±2.43	31.82 ±2.63	42.34 ±2.22	41.88 ±1.76	40.55 ±0.09	40.58 ±0.01
PSB-34	8.65 ±1.72	8.21 ±3.66	12.82 ±3.15	23.66 ±2.43	25.92 ±3.33	35.57 ±2.99	36.93 ±1.21	41.94 ±3.44	41.65 ±2.72	39.77 ±1.5	39.12 ±2.65	39.13 ±0.87	31.91 ±1.41
PSB-37	14.22 ±3.19	31.46 ±1.53	39.81 ±2.42	44.51 ±1.53	44.85 ±1.02	45.51 ±2.03	42.36 ±1.07	41.29 ±2.1	42.29 ±2.17	38.4 ±1.32	39.42 ±1.19	36.2 ±2.39	35.25 ±1.31
PSB-41	10.12 ±1.34	21.25 ±1.43	30.38 ±2.5	30.78 ±1.3	33.70 ±0.96	36.60 ±2.35	33.48 ±1.54	32.29 ±2.31	32.99 ±0.82	31.85 ±1.63	31.3 ±1.45	31.15 ±0.91	31.74 ±1.0
PSB-44	14.36 ±1.54	28.67 ±2.11	35.08 ±1.22	36.62 ±2.64	37.98 ±2.64	38.22 ±1.65	42.66 ±2.44	42.73 ±2.76	41.77 ±3.54	41.71 ±3.34	34.67 ±2.54	33.66 ±1.32	32.94 ±1.55
PSB-48	13.51 ±1.32	23.95 ±1.22	31.06 ±1.64	33.74 ±1.88	36.75 ±3.76	32.44 ±0.91	32.99 ±0.98	31.61 ±1.75	31.85 ±1.65	31.13 ±1.43	31.39 ±1.54	26.45 ±1.08	23.44 ±0.97

Table 25: pH changes during P-solubilization by phosphate solubilising bacterial isolates grown in NBRIP broth medium during 264 h of incubation

Isolates	0 h	6 h	12 h	24 h	48 h	72 h	96 h	120 h	144 h	168 h	192 h	216 h	240 h	264 h
PSB-12	7.0	6.6 ± 0.23	5.9 ± 0.23	5.17 ± 0.27	4.75 ± 0.26	4.6 ± 0.33	4.67 ± 0.34	4.60 ± 0.39	4.50 ± 0.53	4.4 ± 0.33	4.4 ± 0.26	4.50 ± 0.27	4.48 ± 0.37	4.4 ± 0.26
PSB-15	7.0	6.8 ± 0.13	6.3 ± 0.19	5.66 ± 0.32	4.71 ± 0.24	4.72 ± 0.21	4.53 ± 0.23	4.3 ± 0.31	4.38 ± 0.33	4.29 ± 0.33	4.5 ± 0.39	4.56 ± 0.33	4.4 ± 0.37	4.7 ± 0.29
PSB-16	7.0	6.9 ± 0.23	6.6 ± 0.27	5.76 ± 0.33	4.41 ± 0.36	4.43 ± 0.16	3.85 ± 0.27	3.92 ± 0.36	3.89 ± 0.42	4.55 ± 0.32	4.6 ± 0.37	4.5 ± 0.23	4.4 ± 0.26	4.6 ± 0.37
PSB-18	7.0	6.7 ± 0.37	6.1 ± 0.36	5.2 ± 0.33	4.27 ± 0.26	4.43 ± 0.35	3.56 ± 0.47	3.89 ± 0.23	3.95 ± 0.14	4.2 ± 0.36	4.4 ± 0.37	4.7 ± 0.26	4.8 ± 0.13	5.1 ± 0.26
PSB-21	7.0	6.43 ± 0.5	6.48 ± 0.16	6.34 ± 0.19	6.31 ± 0.17	5.58 ± 0.15	5.48 ± 0.17	5.62 ± 0.38	5.1 ± 0.18	4.98 ± 0.26	4.8 ± 0.32	4.48 ± 0.37	4.42 ± 0.32	4.5 ± 0.46
PSB-26	7.0	6.4 ± 0.37	5.2 ± 0.23	4.43 ± 0.33	4.08 ± 0.25	3.83 ± 0.29	3.59 ± 0.49	3.39 ± 0.39	3.23 ± 0.26	3.41 ± 0.32	3.9 ± 0.32	4.41 ± 0.23	4.98 ± 0.37	5.4 ± 0.33
PSB-27	7.0	6.8 ± 0.23	6.5 ± 0.27	5.76 ± 0.37	5.32 ± 0.29	5.03 ± 0.38	5.09 ± 0.29	4.83 ± 0.18	4.88 ± 0.27	4.2 ± 0.26	4.1 ± 0.37	3.9 ± 0.27	4.1 ± 0.56	4.3 ± 0.37
PSB-28	7.0	6.1 ± 0.17	5.2 ± 0.16	4.88 ± 0.27	4.81 ± 0.38	4.96 ± 0.43	4.88 ± 0.33	4.34 ± 0.39	4.23 ± 0.29	4.05 ± 0.19	4.2 ± 0.33	5.4 ± 0.37	5.5 ± 0.47	5.6 ± 0.32
PSB-29	7.0	6.6 ± 0.26	6.5 ± 0.27	4.87 ± 0.18	4.67 ± 0.39	4.22 ± 0.19	4.20 ± 0.29	4.37 ± 0.49	4.15 ± 0.19	3.98 ± 0.34	3.4 ± 0.47	3.4 ± 0.47	3.76 ± 0.26	3.8 ± 0.32
PSB-34	7.0	6.8 ± 0.17	6.1 ± 0.13	5.62 ± 0.15	5.89 ± 0.59	5.11 ± 0.17	4.84 ± 0.37	4.79 ± 0.26	4.01 ± 0.29	4.1 ± 0.33	4.4 ± 0.32	5.1 ± 0.16	5.3 ± 0.27	5.1 ± 0.33
PSB-37	7.0	5.3 ± 0.36	4.1 ± 0.27	3.7 ± 0.28	3.15 ± 0.18	3.29 ± 0.19	3.42 ± 0.17	3.46 ± 0.27	3.89 ± 0.19	4.0 ± 0.26	4.5 ± 0.53	4.5 ± 0.26	4.9 ± 0.47	4.9 ± 0.27
PSB-41	7.0	5.5 ± 0.17	4.5 ± 0.22	4.70 ± 0.27	4.48 ± 0.13	4.42 ± 0.17	4.33 ± 0.34	4.39 ± 0.57	4.40 ± 0.23	4.3 ± 0.37	4.4 ± 0.13	4.4 ± 0.33	4.33 ± 0.36	4.2 ± 0.16
PSB-44	7.0	6.5 ± 0.13	5.5 ± 0.27	4.79 ± 0.28	4.58 ± 0.27	4.59 ± 0.29	4.68 ± 0.39	3.75 ± 0.51	3.95 ± 0.37	4.3 ± 0.32	3.6 ± 0.37	4.15 ± 0.46	4.28 ± 0.33	4.2 ± 0.27
PSB-48	7.0	6.3 ± 0.27	5.3 ± 0.17	4.75 ± 0.14	4.61 ± 0.32	4.10 ± 0.22	4.46 ± 0.29	4.91 ± 0.55	4.61 ± 0.21	4.89 ± 0.37	4.7 ± 0.13	4.73 ± 0.47	5.15 ± 0.37	5.4 ± 0.33

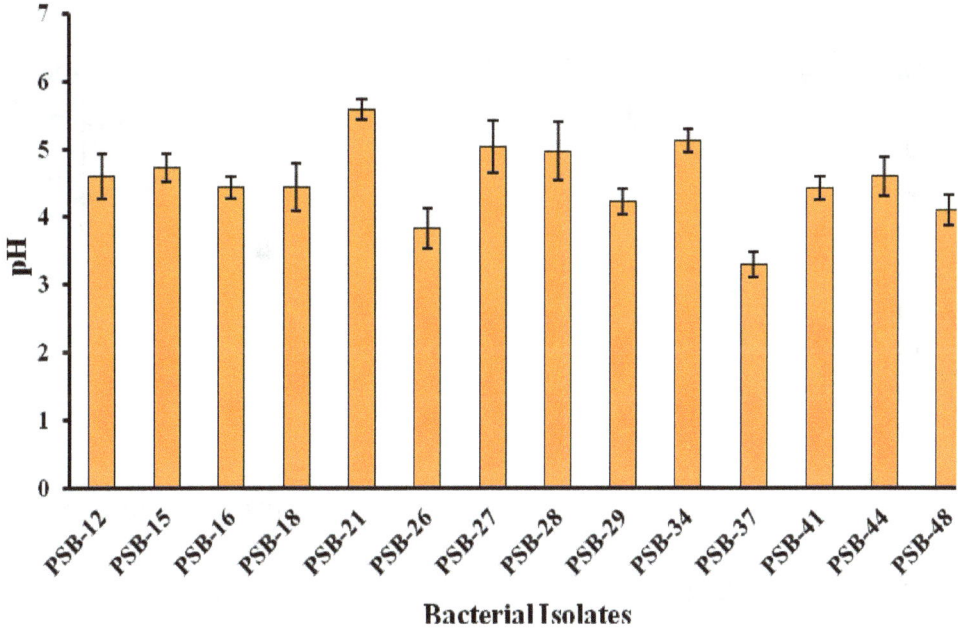

Fig. 39: Changes of pH of the medium by 14 selected phosphate solubilising bacterial isolates from mangrove soilof Mahanadi delta

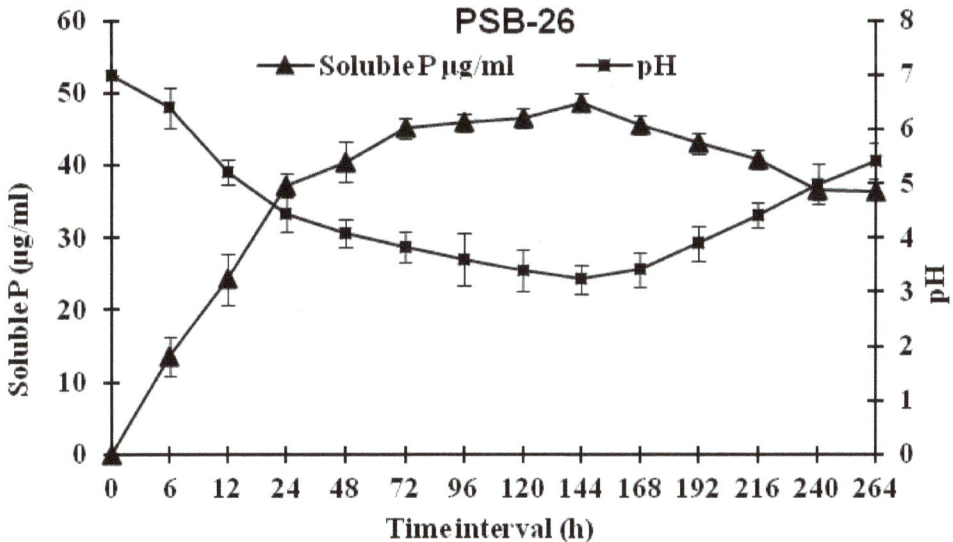

Fig. 40: Tricalcium phosphate solubilisation and changes of pH of the medium by the bacterial isolate PSB-26.

PSB-37

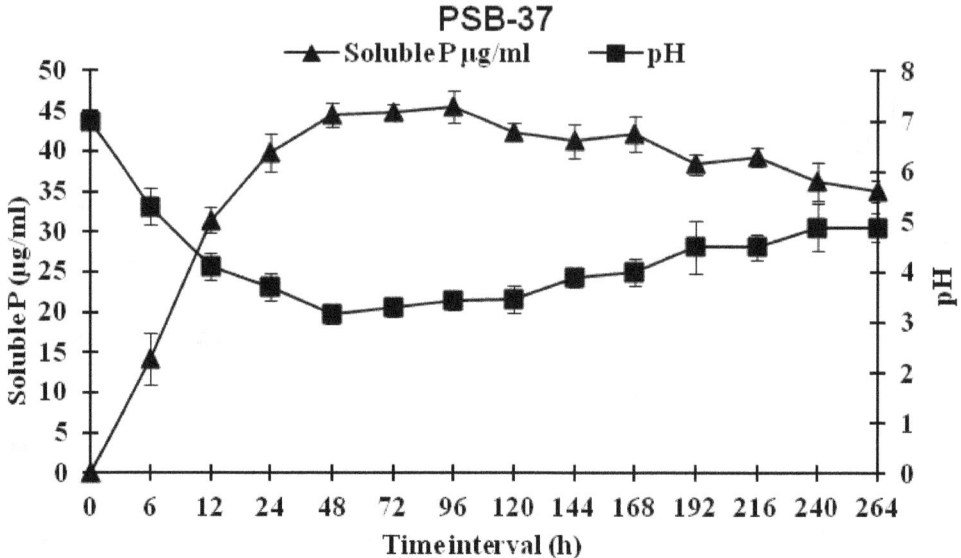

Fig. 41: Tricalcium phosphate solubilisation and changes of pH of the medium by the bacterial isolate PSB-37

4.2.1.5. Detection of organic acids through HPLC analysis

As two bacterial isolates, PSB-26 and PSB-37 showed maximum phosphate solubilisation and reduction in pH, hence, organic acid analysis was carried out taking the broth culture of these two bacterial isolates only. For the analysis of organic acids, strains were inoculated in 250 ml conical flask containing 50 ml NBRIP broth. The flasks were incubated at 37 °C temperature in an orbital shaker at 180 rpm for 96 h. One ml of incubated sample was centrifuged at 10,000 g (Mikro-200, Hettich Zentrifugen, Germany) for 15 min and was filtered through 0.2 μm nylon membranes (Pall India Pvt. Ltd.) to obtain cell free culture supernatant. From the cell free supernatant 20 μl was injected to HPLC (LC-10AT, Shimadzu). The organic acid separation was carried out on ion exclusion column, Aminex® HPX-87H, 300 mm x 7.8 mm (Bio-Rad Laboratories, Inc.) with 0.008 M H_2SO_4 as mobile phase at a constant flow rate of 0.6 ml/min and at operating temperature of 30 °C. Retention time of each signal was recorded at a wavelength of 210 nm (SPD 10A, Shimadzu) and compared with organic acid analysis standard kit (Bio-Rad Laboratories, Inc.) whereas lactic acid and propionic acid of Sigma, USA was injected separately.

Analysis of organic acid in the culture broth was performed by HPLC after 96 h of incubation. Six organic acids were taken as standard and their retention time were summerised in Fig. 42 and table 26. Five different organic acids such as citric acid, oxalic acid, malic acid, succinic acid and acetic acids were produced from the culture medium of the isolate PSB-26 (Fig. 44). All the organic acids produced by the isolates were also quantified and presented in Table 27. Among the five different organic acids, oxalic acid was found to be secreted maximum (289 mg/l) by the isolate PSB-26.

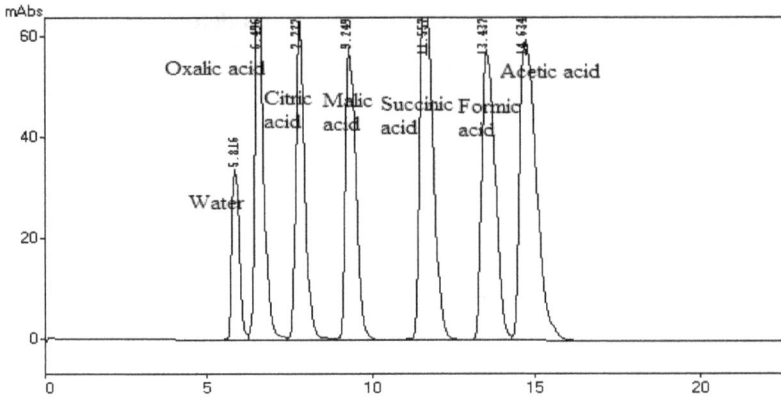

Fig. 42: Picks shown by six organic acid standards analysed through HPLC

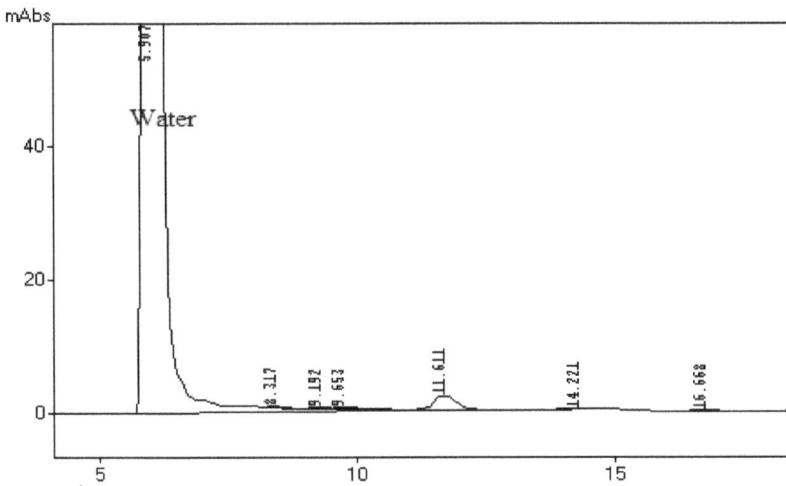

Fig. 43: picks shown by control runned in HPLC

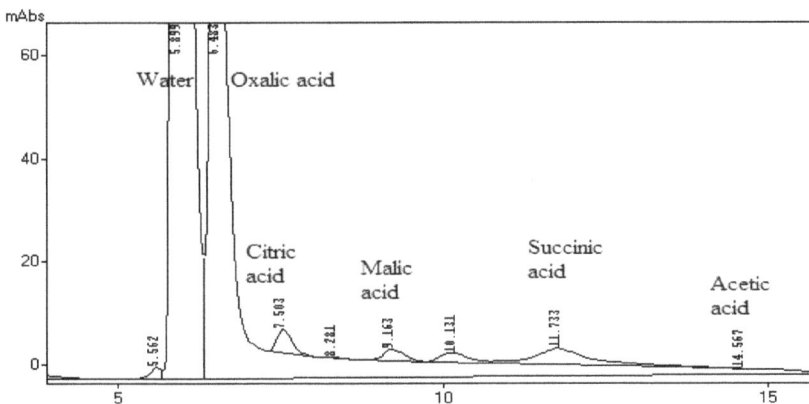

Fig. 44: Organic acids detectection through HPLC by the bacterial isolate, PSB-26

Table 26: Retention time of standard organic acids detected through HPLC

Organic Acid	Retention Time
Oxalic Acid	6.5
Citric Acid	7.7
Malic Acid	9.2
Lactic Acid	10.2
Succinic Acid	11.4
Formic Acid	13.4
Acetic Acid	14.5

Table 27: Quantification of organic acids produced by bacterial isolate, PSB-26

Organic Acid produced	Amount (mg/l)
Oxalic acid	289
Citric acid	0.2
Malic acid	0.3
Succinic acid	0.5
Acetic acid	0.4

Similarly four different organic acids such as malic acid, lactic acid, acetic acid and propionic acid were detected from PSB-37 (Fig. 45). All the organic acids produced by the isolate, PSB-37 was quantified and presented in **Table 28**. Highest acid produced by the isolate, PSB-37 was found to be lactic acid (599.5 mg/l) followed by malic acid (237 mg/l).

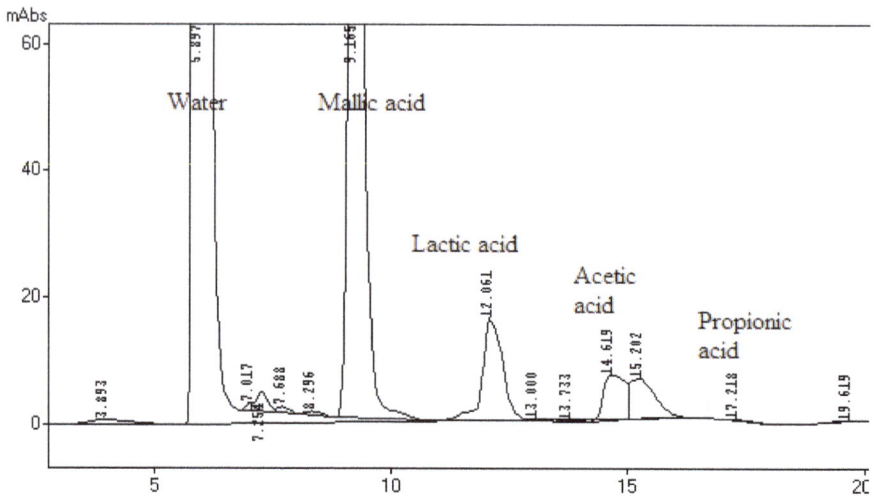

Fig. 45: Organic acid detected through HPLC by bacterial isolate PSB, 37

Table 28: Quantification of organic acids produced by bacterial isolate, PSB-37

Organic acid produced	Amount (mg/l)
Lactic acid	599.5
Acetic acid	5.0
Malic acid	237.0
Propionic acid	8.0

4.2.1.6. Evaluation of alkaline and acid phosphatase production

Evaluation of alkaline and acid phsphatase production was studied for all the seleted fourteen efficient phosphate solubilising bacteria (PSB-12, PSB-15, PSB-16, PSB-18, PSB-21, PSB-26, PSB-27, PSB-28, PSB-29, PSB-34, PSB-37, PSB-41, PSB-44 and PSB-48). For extraction of enzyme, experiment was carried out by taking 100 ml of sterilised NBRIP broth in 250ml conical flask and inoculated with the selected culture. The inoculated flasks were incubated at 37 °C. The samples were drawn and centrifuged at 10,000 rpm for 10 minutes at 4 °C and the cell free supernatant fluid (as crude enzyme) was assayed for crude phosphatase activity. Modified universal buffer having pH of 6.5 and 11.0 were used for determination of acid phosphatase (AcPase) and alkaline phosphatase (AlPase) activity, respectively as described by Tabatabai and Bremner (1969).

It has been observed that all the bacterial isolate showed increase in phosphatase activity upto a certain incubation period and after which it decreased gradually. From the comparative studies of each bacterial isolates it is found that the alkaline phosphatase production by different bacteria ranged between 7.53- 71.531 U/ml (Table-29) where as acid phosphatase production of the bacteria was in the range of 2.55-76.8 U/ml (Table-30). Among 14 isolates, PSB-26 showed highest alkaline phophatase production at 48 hour of incubation where as the bacterial isolate PSB-37 produced highest acid phosphatase production (76.8 U/ml) during 48 hour of incubation.

4.2.1.7. Selection of most efficient strain among the selected 14 bacterial isolates

Based on their comparative phosphate solubilising efficiency, quantitative P solubilisation and phosphatase production ability (**Table 31**), two bacterial isolates such as, PSB-26 and PSB-37 were selected as the most efficient strains and further work were carried out with these two selected bacterial isolates.

Table 29: Alkaline phosphatase production (U/ml) by bacterial isolates from mangrove soils of Mahanadi delta, Odisha incubated for 192 h.

Isolates	0	6 h	12 h	24 h	48 h	72 h	96 h	120 h	144 h	168 h	192 h
PSB-12	0	7.533 ± 1.31	33.723 ± 1.43	30.957 ± 1.52	32.446 ± 1.48	30.957 ± 1.03	29.787 ± 1.09	30.425 ± 1.09	26.914 ± 1.65	24.68 ± 1.42	24.616 ± 1.98
PSB-15	0	10.744 ± 1.43	46.382 ± 1.53	46.808 ± 2.33	44.68 ± 1.67	36.17 ± 1.83	28.51 ± 1.65	26.595 ± 1.76	26.489 ± 1.53	27.231 ± 1.75	10.106 ± 0.78
PSB-16	0	9.787 ± 1.47	29.999 ± 1.87	37.872 ± 1.83	35.319 ± 1.22	40.638 ± 1.72	37.446 ± 1.87	36.914 ± 1.54	36.808 ± 1.76	34.255 ± 1.98	10.382 ± 1.53
PSB-18	0	10.957 ± 1.09	40.851 ± 2.11	44.893 ± 1.43	44.362 ± 1.32	29.787 ± 1.62	37.234 ± 1.65	31.914 ± 0.87	27.553 ± 1.59	30.106 ± 1.65	22.553 ± 1.54
PSB-21	0	27.127 ± 2.08	33.723 ± 1.76	49.787 ± 1.30	60.425 ± 2.35	28.404 ± 1.39	24.148 ± 1.54	42.659 ± 0.76	31.382 ± 1.79	31.063 ± 2.13	10.319 ± 1.52
PSB-26	0	14.754 ± 1.76	43.191 ± 1.54	59.893 ± 1.64	71.531 ± 1.88	62.51 ± 1.39	50.702 ± 1.83	46.382 ± 0.65	27.872 ± 1.34	27.872 ± 1.87	15.638 ± 2.47
PSB-27	0	9.999 ± 0.99	40.531 ± 1.47	32.446 ± 1.39	41.17 ± 1.99	35.319 ± 1.64	30.957 ± 2.43	29.468 ± 1.54	29.574 ± 1.65	28.617 ± 2.87	20.021 ± 2.44
PSB-28	0	15.638 ± 1.55	32.978 ± 1.44	55.957 ± 1.39	66.276 ± 1.99	58.829 ± 0.54	37.756 ± 1.43	34.148 ± 1.33	36.489 ± 1.48	35.957 ± 1.65	16.914 ± 2.76
PSB-29	0	12.533 ± 1.43	45.744 ± 2.69	45.531 ± 1.43	42.872 ± 1.54	26.072 ± 0.87	28.297 ± 2.76	26.382 ± 1.54	26.914 ± 1.53	28.297 ± 1.84	10.063 ± 2.52

Isolates	0	6 h	12 h	24 h	48 h	72 h	96 h	120 h	144 h	168 h	192 h
PSB-34	0	24.042 ± 1.32	42.978 ± 1.65	48.51 ± 0.99	39.361 ± 1.98	27.978 ± 0.98	28.51 ± 1.43	28.829 ± 1.42	27.553 ± 1.29	26.914 ± 1.45	14.787 ± 1.66
PSB-37	0	11.489 ± 0.87	38.404 ± 2.31	38.51 ± 0.62	36.382 ± 2.76	29.787 ± 1.87	29.747 ± 1.43	26.17 ± 1.94	26.382 ± 1.20	25.212 ± 1.42	14.468 ± 1.54
PSB-41	0	9.574 ± 1.51	36.702 ± 1.09	42.234 ± 0.73	36.17 ± 2.54	30.212 ± 1.67	49.574 ± 1.33	38.723 ± 1.65	28.191 ± 1.33	28.404 ± 1.85	14.148 ± 1.32
PSB-44	0	8.297 ± 1.32	33.404 ± 1.89	42.234 ± 1.53	40.531 ± 2.31	47.127 ± 1.65	45.638 ± 1.87	37.978 ±1.54	30.425 ± 1.32	29.787 ± 0.94	24.893 ± 1.52
PSB-48	0	9.574 ± 0.84	33.297 ± 1.48	34.468 ± 0.73	38.085 ± 3.1	28.829 ± 1.76	26.595 ± 1.54	29.042 ± 1.64	29.68 ± 1.21	28.936 ± 1.53	16.382 ± 1.72

Table 30: Acid Pase production (U/ml) by bacterial isolates from mangrove soils of Mahanadi delta, Odisha incubated for 192 h

Time (h) / Isolates	0	6	12	24	48	72	96	120	144	168	192
PSB-12	0	5.425 ± 1.64	38.404 ± 1.76	38.829 ± 1.54	42.659 ± 1.65	36.808 ± 1.09	36.17 ± 1.29	33.191 ± 1.28	30.106 ± 1.29	30.233 ± 1.94	29.987 ± 1.85
PSB-15	0	24.68 ± 0.76	54.754 ± 1.63	58.191 ± 1.78	64.787 ± 0.99	42.234 ± 1.22	47.021 ± 0.59	40.425 ± 1.38	32.234 ± 1.27	31.489 ± 1.55	10.531 ± 2.33
PSB-16	0	14.99 ± 0.87	33.829 ± 1.89	47.127 ± 1.65	46.382 ± 1.87	47.021 ± 1.27	48.723 ± 0.39	49.574 ± 1.94	52.021 ± 1.98	36.489 ± 0.95	11.17 ± 2.52
PSB-18	0	11.276 ± 1.11	24.574 ± 1.54	33.51 ± 1.42	38.085 ± 1.09	65.319 ± 1.27	42.34 ± 1.82	40.638 ± 0.67	38.829 ± 2.31	35.531 ± 0.83	3.085 ± 1.52

Time (h) / Isolates	0	6	12	24	48	72	96	120	144	168	192
PSB-21	0	18.297 ± 0.65	34.68 ± 1.69	34.999 ± 1.42	36.702 ± 2.14	43.617 ± 1.89	44.148 ± 1.20	45.957 ± 1.76	36.063 ± 1.29	26.595 ± 1.29	15.531 ± 1.62
PSB-26	0	16.068 ± 1.56	27.234 ± 1.98	45.063 ± 1.72	38.021 ± 2.11	41.957 ± 1.64	40.425 ± 1.29	36.808 ± 1.25	36.595 ± 1.49	33.085 ± 1.39	12.978 ± 1.69
PSB-27	0	10.744 ± 1.87	31.063 ± 1.54	51.702 ± 1.1	42.659 ± 1.83	43.723 ± 1.33	37.553 ± 1.36	35.957 ± 1.20	30.744 ± 1.52	29.574 ± 1.33	9.893 ± 1.33
PSB-28	0	8.191 ± 1.63	38.191 ± 1.90	37.659 ± 0.92	38.51 ± 1.22	43.085 ± 1.77	39.361 ± 1.33	39.787 ± 0.73	39.957 ± 1.92	39.787 ± 1.39	10.425 ± 0.97
PSB-29	0	8.829 ± 0.74	34.042 ± 2.31	44.574 ± 0.98	45.851 ± 0.52	52.872 ± 1.43	43.723 ± 1.47	43.51 ± 0.83	32.659 ± 1.76	30.319 ± 1.93	22.446 ± 0.93
PSB-34	0	10.851 ± 1.99	37.659 ± 2.09	40.106 ± 1.78	40.851 ± 1.73	55.319 ± 1.32	38.936 ± 1.58	36.063 ± 1.28	25.212 ± 1.92	24.574 ± 1.58	17.234 ± 1.73
PSB-37	0	11.914 ± 0.54	29.787 ± 1.54	49.148 ± 1.32	76.808 ± 2.99	67.021 ± 1.83	55.331 ± 1.59	42.304 ± 1.78	37.127 ± 1.83	33.297 ± 1.66	13.297 ± 1.37
PSB-41	0	8.829 ± 1.69	44.148 ± 1.33	64.255 ± 3.33	53.829 ± 2.17	41.702 ± 1.92	41.914 ± 1.29	32.021 ± 1.77	30.531 ± 1.77	28.191 ± 0.75	2.553 ± 1.27
PSB-44	0	6.914 ± 1.93	32.872 ± 1.31	60.638 ± 1.65	60.531 ± 1.49	62.978 ± 2.87	57.127 ± 1.27	56.808 ± 1.93	42.304 ± 0.89	39.255 ± 1.67	9.787 ± 1.22
PSB-48	0	24.574 ± 0.99	37.021 ± 1.65	43.404 ± 1.43	52.34 ± 1.43	35.957 ± 1.53	35.425 ± 1.25	34.999 ± 0.92	32.021 ± 0.95	30.319 ± 1.50	15.531 ± 1.72

Table 31: Comparative Phosphate solubilisation, pH change and phosphatase production by selected fourteen bacterial isolates

Bacterial Isolates	Soluble P (µg/ml)	pH change	AlPase (U/ml)	AcPase (U/ml)
PSB-12	38.69 ± 1.23	4.2 ± 1.43	33.72 ± 1.43	42.65 ±1.65
PSB-15	36.69 ± 2.75	4.29 ± 0.33	46.8 ± 2.33	64.78 ± 0.99
PSB-16	38.0 ± 2.71	3.85 ± 0.27	40.63 ± 1.72	52.09 ± 1.43
PSB-18	39.79 ± 3.59	3.56 ± 0.47	44.89 ± 1.43	65.31 ± 1.27
PSB-21	33.14 ± 1.75	4.42 ± 0.32	60.42 ± 2.35	45.95 ± 1.76
PSB-26	48.70 ± 3.54	3.23 ± 0.26	71.53 ± 1.88	45.06 ± 1.72
PSB-27	33.94 ±1.33	3.9 ± 0.27	41.1 ± 1.99	51.7 ±1.33
PSB-28	37.35 ± 3.88	4.05 ± 0.19	66.27 ± 1.99	43.08 ± 1.77
PSB-29	42.34 ± 1.42	3.4 ± 0.47	45.74 ± 2.69	52.87 ± 1.43
PSB-34	41.94 ± 3.44	4.01 ± 0.29	48.51 ± 0.99	55.31 ± 1.32
PSB-37	44.84 ± 3.19	3.15 ± 0.18	38.51 ± 0.62	76.8 ± 2.99
PSB-41	36.60 ± 2.35	4.21 ± 0.16	49.57 ± 1.33	64.25 ± 3.33
PSB-44	42.73 ± 3.54	3.75 ± 0.51	47.12 ± 1.65	62.97 ± 2.87
PSB-48	36.75 ± 3.76	4.1 ± 0.22	38.08 ± 3.1	52.34 ± 1.43

4.2.1.8. Optimization of alkaline phosphatase

To obtain the maximum alkaline phosphatase production the bacterial strain PSB-26 was inoculated in different growth condition such as different pH, temperature, agitation, carbon sources and nitrogen sources. The optimum alkaline phosphatase activity was observed at 48 h of incubation (71.531 U/ml, p<0.001), pH 9.0 (82.91 U/ml, p<0.001) temperature of 45 °C (84.99 U/ml, p<0.001) an agitation rate of 100 rpm (87.085 U/ml) and with glucose as a original carbon source (87.66 U/ml, p<0.001) and ammonium sulfate as a original nitrogen source (88.66 U/ml, p<0.001). Under optimized sets of conditions, maximum alkaline phosphatase activity of 93.7 U/ml was observed (Fig. 46a-e).

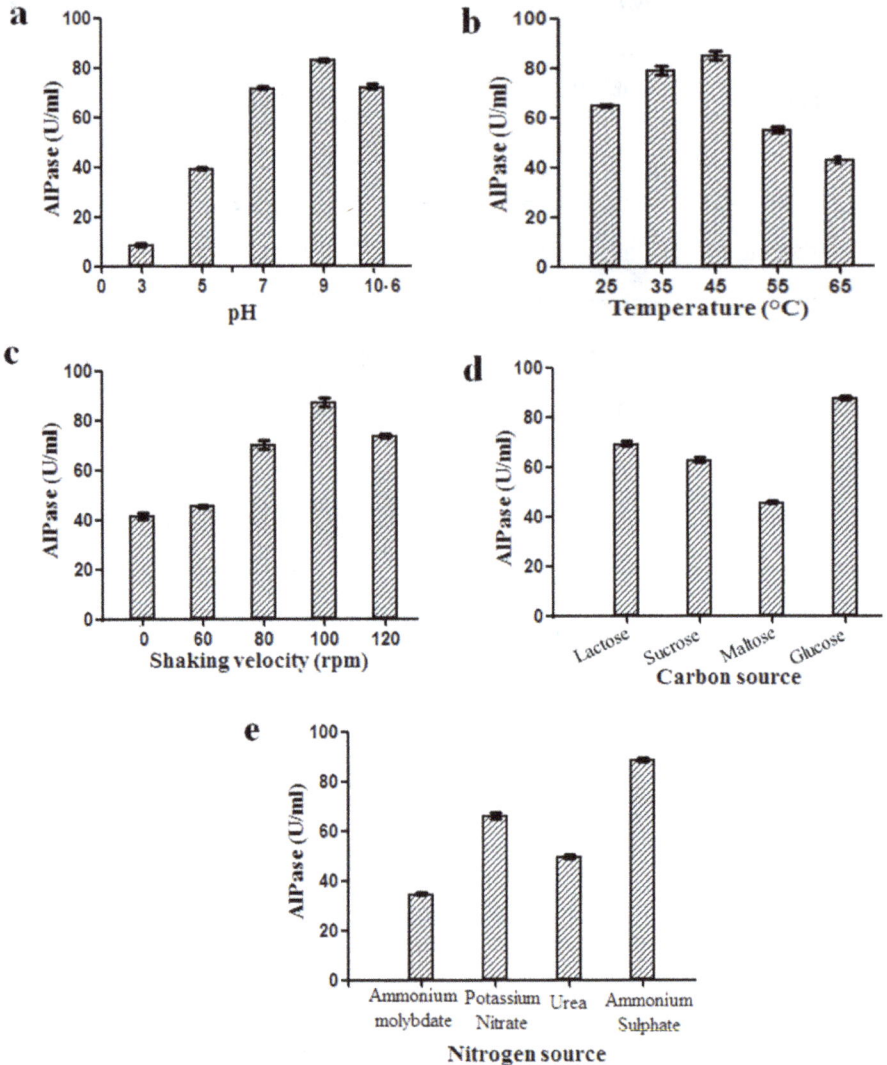

Fig. 46(a-e). Effect of different growth parameters such as (a) pH, (b) Temperature, (c) Shaking velocity, (d) Carbon sources and (e) Nitrogen sources on crude alkaline phosphatase production by the isolate, PSB-26

4.2.1.9. Partial purification of alkaline phosphatase

Partial purification of alkaline and acid phosphatase was carried out by ammonium sulphate precipitation followed by dialysis. To the chilled cell free culture fluid, solid ammonium sulphate was added with gentle stirring to 30% saturation and incubated at 4°C for 2 h. It was then centrifuged at 10000 rpm for 10 min at 4°C and the precipitate was discarded. More ammonium sulphate was added to 70% saturation. The precipitate was collected by centrifugation after two h incubation at 4°C and dissolved in 0.2M phosphate buffer (pH 7). This solution was dialyzed overnight against the same buffer at 4°C. This dialyzed enzyme was used for further studies. Estimation of total protein was performed as per the method of Lowry *et al.*, (1951), taking Bovine serum albumin as standard. Results showed that the partially purified alkaline phosphatase from the bacterial isolate, PSB-26 exhibited protein content of 6 mg/ml with a specific activity of 16.33 U/mg which corresponds to 1.82 fold purification and 41.09% yield (Table 32).

Table 32: Partial purification of AlPase

Isolates PSB-26	Total Volume (ml)	Protein concentration (mg/ml)	Total Protein content (mg)	Phosphatase activity (U/ml)	Total phosphatase Activity (U)	Specific Activity (U/mg)	Fold of Purification	Total Yeild (%)
Crude extract	50	8.0	400	71.53	3576.55	8.94	1.0	100%
70% (NH₄)₂SO₄ precipitation and dialysis	15	6.0	90.0	7.99	1469.85	16.33	1.82	41.09%

4.2.1.10. Characterization of alkaline phosphatase activity

The molecular weight of the partially purified enzyme were carried out by SDS-PAGE, using 5% stacking gel and 10% resolving gel and electrophoresis was performed with a 15 mA fixed current (Laemmli, 1970; Sasirekha *et al.*, 2012). The gel was stained with coomassie brilliant blue R250 and destained with destaining solution (methanol: acetic acid: water: 30: 10: 10) for 8-10 h. The effect of pH on partially purified alkaline activity were determined using different buffers ranging from 0.1M citrate buffer pH 3.0, 0.1M acetate buffer pH 5.0, 0.1M phosphate buffers pH 7.0, 0.1M glycine-sodium hydroxide buffer pH 9 and 10.6 in the reaction mixture. The enzyme activities were measured after incubation for 1h at 37 °C.

Effect of temperature on partially purified alkaline phosphatase activity was determined by incubating the reaction mixture at 10°C intervals of temperatures from 25 to 65 °C for 30 minutes and the enzyme activities were measured.

Effects of substrate concentration on partially purified alkaline phosphatase activities were determined at varying the concentration of p-nitrophenyl phosphate in the reaction mixture from 0.5 mg to 2.5 mg/ml. After incubation at 37 °C, the enzyme activities were determined by measuring the p-nitrophenol formed. After partial purification, the enzyme showed band sizes of approximately 45 kDa, 25 kDa and 17kDa on SDS-PAGE **(Fig. 47a).**

The activity of partially purified alkaline phosphatase was checked with different parametric conditions such as in different pH (3.0-10.6), temperature 25-65 °C and substrate concentration ranged from (0.5-2.5 mg/ml) and the data is presented in Fig. 47(b-d). The maximum alkaline phosphatase activity of the bacterial isolate, PSB-26 was recorded at pH 9.0 (96.53 U/ml, p<0.001).The optimum temperature for enzyme activity was found to be 45°C with maximum enzyme activity of 97.99 U/ml, p<0.001. It has been observed that the enzyme activity increased with increasing in substrate concentration up to 2 mg/ml of p-nitrophenyl phosphate concentration with a maximum activity of 96.51 U/ml, p<0.001 (V_{max} value 213.70±42.43 and K_m value 2.61±0.87) and remained constant thereafter.

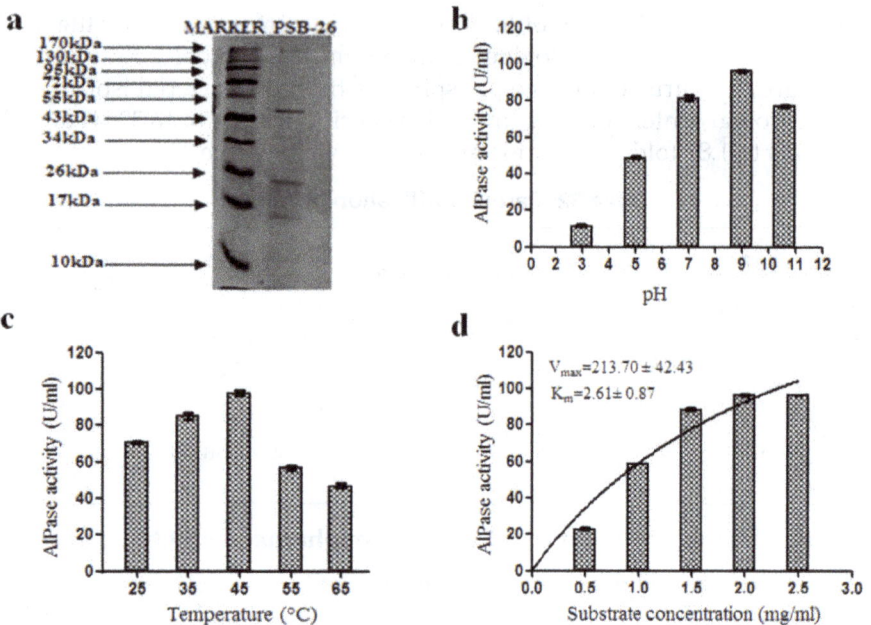

Fig. 47 (a) SDS-PAGE of partially purified Alpase from PSB-26. (Proteins were visualized with Coomassie blue staining. Lanes 1: Molecular weight standards; Lane 2: Cell supernatant after partial purification), (b) effect of pH, (c) Temperature, (d) Substrate concentration on activity of partially purified alkaline phosphatase from PSB-26.

4.2.1.12. Effect of PSB on plant growth promotion

The effect of phosphate solubilizing bacteria on the growth of *A. thaliana* (Col-0) was studied by growing the surface sterilised seeds on a planton tissue culture container (7.5x7.5x10mm) containing 40 ml of Murashige and Skoog (MS) agar medium (Murashige and Skoog, 1962). Three seeds were taken in each planton with three replicas for each observation. The original MS media was kept as the positive control (MS media + sterile seed) where as in the test; the original soluble P source (KH_2PO_4) was replaced with insoluble tricalcium phosphate (TCP) and the modified medium was named as modified MS+TCP media. Overnight grown bacterial culture (10µl) were inoculated with the modified MS+TCP media containing sterilised seeds of *A. thaliana* and considered as test planton. The MS+TCP media inoculated with sterilised seed (modified MS+TCP media + sterile seed) without bacterial inoculation was considered as the negative control. All test plantons, positive control plantons and negative control plantons were incubated at 4°C for 2-3 days for cold stratification. After stratification, all the plantons were placed into the long day (16 hr day and 8 hr night) light cabinet (J.K.G Bioscience pvt.ltd, Ohio) and kept at 22°C and humidity of 65% for 30 days.

Comparative analysis of vegetative and reproductive plant growth patterns i.e., the date of sowing, date of germination, number of plants germinated, number of leaves, length of root, length of shoot and number of flower were recorded in triplicate from germination of seeds up to maturation of the plant. Plant development data were collected and presented in Table 33. Minimum shoot/root ratio was observed in the negative control planton (modified MS+TCP media + sterile seed without bacterial inoculation), which showed very stunted growth (Fig. 48a) in comparison to the test (modified MS+TCP media + sterile seed + 10µl bacterial inoculation) (Fig. 48c) and positive control planton (MS media + sterile seed without bacterial inoculation) (Fig. 48b). Higher plant growth was observed in the planton containing modified MS+TCP media inoculated with bacterial culture (**Fig. 48c**).

Table 33 Effect of *A. faecalis* on plant growth

Plant specification	Number of leafs	Number of flowers	Length of shoot (cm)	Length of root (cm)	Shoot/ Root
MS media + Plant	6 ± 1.41	4 ±1.41	6.76±0.52	3.16±0.11	2.14
(modified MS+TCP media) + Plant	4 ± 0	0	0.42±0.11	0.25±0.05	1.67
(modified MS+TCP media) + plant + 10µl bacteria	7 ± 1.41	30 ±4.24	15.11±1.46	1.64±0.08	9.24

Modified MS + TCP media: MS media in which KH_2PO_4 replaced with TCP

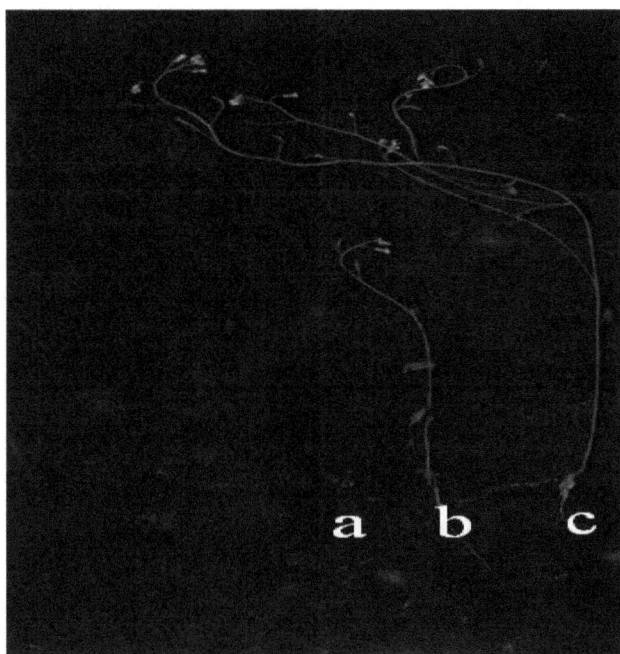

Fig. 48. Effect of *A. faecalis* **on** *A. thaliana* **plant growth promotion. The plantons are (a) modified MSTCP media + plant without bacterial inoculation as negative control, (b) MS media + plant without bacterial inoculation as positive control, (c) modified MSTCP media + plant with 10µl of bacterial inoculation as test planton.**

4.2.1.13. Optimization of acid phosphatase

Similar to alkaline phosphatase to obtain the maximum acid phosphatase production the bacterial culture of PSB-37 was grown in different growth condition such as different pH, temperature, agitation, carbon sources and nitrogen sources and optimum acid phosphatase activity was observed at 48 h of incubation (76.808 U/ml), temperature of 45 °C (77.87 U/ml) (**Fig. 49a**), an agitation rate of 100 rpm (80.40 U/ml) (**Fig. 49b**), pH 5.0 (80.66 U/ml) (**Fig. 49c**) and with glucose as a original carbon source (80.6 U/ml) (**Fig. 49d**) and ammonium sulfate as a original nitrogen source (80.92 U/ml) (**Fig. 49e**). Characterization of the partially purified acid phosphatase showed maximum activity at pH 5.0 (85.6 U/ml), temperature of 45 °C (97.87 U/ml) and substrate concentration of 2.5 mg/ml (92.7 U/ml).

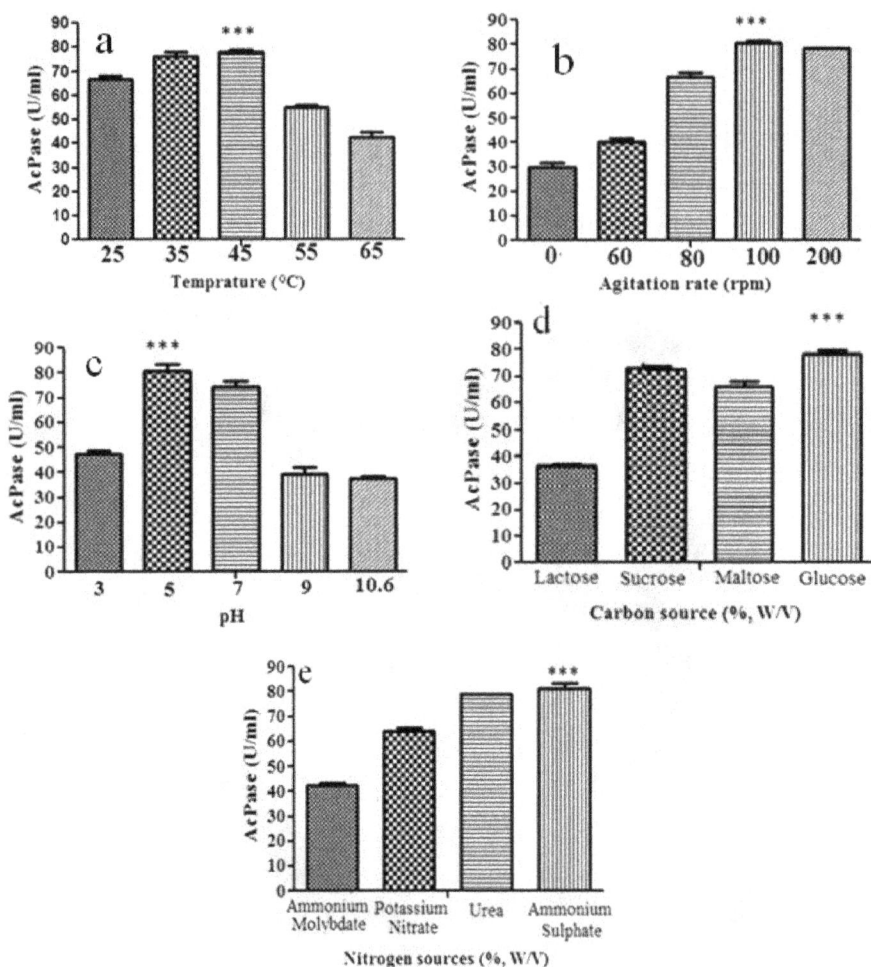

Fig. 49 Effect of different growth parameters such as (a) pH, (b) Temperature, (c) Shaking velocity, (d) Carbon sources and (e) Nitrogen sources on crude acid phosphatase production by the isolate, PSB-37

4.2.1.14. Partial purification & Characterization of acid phosphatase

Partial purification of crude acid phosphatase was carried out by ammonium sulfate precipitation up to 70% saturation followed by overnight dialysis Sasirekha *et al.*, 2012). Quantification of protein content of crude and partially purified phosphatase was carried out following the method of Lowry *et al.*, (1951) with Bovine serum albumin as a standard (Sigma, Germany). The impact of parameters, such as pH from 3-10.6 (by adjusting the pH of the buffer), temperature (at 10 °C intervals) from 25-65 °C and different substrate concentrations (p-nitrophenyl phosphate) from 0.5 mg to 2.5 mg ml^{-1} were studied in triplicate for characterization of partially purified acid phosphatase activity following the method of Tabatabai and Bremner (1969). Partial purification of acid phosphatase was performed by 70% ammonium sulphate precipitation followed by over night dialysis and results obtained are presented in Table 34. Acid phosphatase from the bacterial isolate, PSB-37 could be purified upto 1.5 fold with 33.68% yield and specific activity of 9.58 U /mg.

Table 34: Partial purification of acid phosphatase

Bacterial isolate	Total Volume (ml)	Protein Conc. (mg/ml)	Total Protein Conc. (mg)	AcPase Activity (U/ml)	Total AcPase Activity (U)	Specific Activity (U/mg)	Fold of Purification	Total Yield (%)
Culture extract of PSB-37	50.0	12.0	600	76.8	3840	6.4	1.0	100%
70% (NH$_4$)$_2$SO$_4$ precipitation and dialysis	15.0	9.0	135	86.23	1293.45	9.58	1.5	33.68%

The effect of pH on partially purified phosphatase activity was determined by assaying the enzyme activity in buffers having different pH, ranged from pH 3.0 to 10.6. The maximum activity of acid phosphatase was recorded (85.6 U/ ml, $p < 0.0001$) at pH 5.0 (Fig. 50a). Enzyme activity was also greatly influenced by the incubation temperature when tested over a wide range of temperatures. The optimum temperature for acid phosphatase activity was found to be 45°C, with an enzyme activity of 97.87 U/ml ($p <0.0001$). Temperatures higher than 45°C resulted in a reduction in acid phosphatase activity probably due to the denaturation of the enzyme (Fig. 50b). The effect of acid phosphatase activity by the isolate PSB-37 was studied over a wide range (0.5–2.5 mg/ml) of substrate concentration. It has been observed that the enzyme activity increased with an increase in substrate concentration, with a maximum activity of 92.7 U/ml ($p < 0.0001$) at 2.5 mg/ml (Fig. 50c).

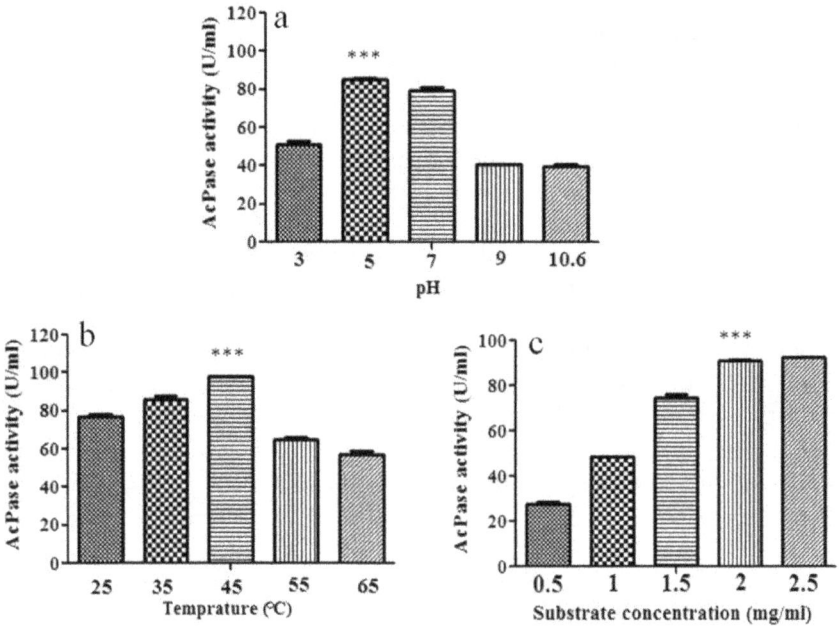

Fig. 50 Effect of pH(a),Temperature(b), and Substrate concentration(c) on activity of partially purified acidphosphatase from PSB-37.

4.3. Cellulose Degrading Bacteria from Mangroves of Mahanadi Delta.

Cellulases are enzymes capable of degrading cellulose, a carbohydrate found as a structural substance of the plant world (Khan and Kumar, 2012). Cellulose found in the cell wall of plant is commonly degraded by the hydrolytic action of a multicomponent enzyme system of cellulase which represents the key step for biomass conversion (Sadhu *et al.*, 2013). The enzymatic hydrolysis of cellulosic biomass brought about by synergistic action of cellobiohydrolase or exoglucanase (E.C. 3.2.1.91), endoglucanase or carboxymethylcellulase (E.C. 3.2.1.4), and cellobiase or β-glucosidase (E.C. 3.2.1.21) (Sadhu *et al.*, 2013). In recent years, there has been growing interest in microbial cellulases that are of commercial importance. Cellulases contribute to 8% of the worldwide industrial enzyme demands and the demand increases gradually (Costa *et al.*, 2008). Apart from degradation of cellulose, the enzyme, cellulases have been used in different industries with prominence to the treatment of agricultural and industrial wastes (Kuhad *et al.*, 2011). Researchers have been focused in cellulases because of their applications in industries of starch processing, grain alcohol fermentation, malting and brewing, extraction of fruit and vegetable juices, pulp and paper industry, and textile industry (Gao *et al.*, 2008). A potential challenging area where cellulases would have a central role is the bioconversion of renewable cellulosic biomass to commodity chemicals including biofuel product. The major constrains for industrial application of this enzyme is the high cost of production and low yields. Therefore, investigations are under taken to find out the efficient cellulose hydrolyzing microbial strains from different environments for production of cellulase, utilizing inexpensive substrates. Availability of highly active cellulase would be of great significance and hence requires selection and improvement of suitable strains for enzyme production and development of fermentation technology for producing them in large quantity (Tolan and Foody, 1999). Many workers have purified and characterized cellulases isolated from different microorganisms and environments namely cellulase from *Thermomonospora* sp. from self-heating compost from the Barabanki district of Uttar Pradesh, India (George *et al.*, 2001), *Cellulomonas* sp. YJ5 from the soil of Northern Taiwan (Yin *et al.*, 2010), *Melanocarpus* sp. MTCC 3922 from composting soil (Kaur *et al.*, 2007), *Pseudomonas fluorescens* (Bakare *et al.*, 2005), *Bacillus* sp. from hot spring (Acharya and Chaudhury 2011; Ashabil *et al.*, 2011).

Mangrove habitats are considered to be a unique niche for the isolation of cellulose-degrading microorganisms due to occurrence of rich lignocellulosic biomass. Thick organic matter mixed with sediment makes mangrove anaerobic except the sediment surface. In such anaerobic environments, decomposition of cellulose such as mangrove leaves and woods are brought about by complex communities of interacting microorganisms by hydrolyzing the β-1,4-glycosidic linkages of cellulose (Alongi 1989). Biodegradation of β-1,4-glycosidic linkages of cellulosic biomass is usually occurred by enzymes such as cellulase produced by numerous microorganisms. Cellulase can be produced by different microorganisms including fungi, bacteria, or actinomycetes. However, bacteria serve as a novel source of cellulases due to their higher growth rate, more complex glycoside hydrolases providing synergy with higher potency because of organismal diversity of extreme niches (Sadhu *et al.*, 2013). Thus cellulose availability forms the basis of many microbial interactions in mangroves (Holguin *et al.*, 2001) and the cellulase enzyme produced by these microorganisms have attracted much interest because of their various applications. Considering the importance and application of the cellulase, cellulose-producing bacteria were isolated from mangrove soils of Mahanadi delta, Odisha, India, for better cellulase production. Further, it was aimed to purify and characterize the crude cellulase enzyme to understand their biotechnological potential.

4.3.1. Isolation of cellulose degrading bacteria

Cellulose degrading bacteria were isolated from, mangrove soil sample of Mahanadi delta Odisha, soils using CMC agar medium(g/l) (carboxymethylcellulose (CMC), 10; tryptone, 2; KH_2PO_4, 4; Na_2HPO_4, 4; $MgSO_4.7H_2O$, 0.2; $CaCl_2.2H_2O$, 0.001; $FeSO_4.7H_2O$, 0.004; Agar, 15, and pH adjusted to 7.0). flooded with 0.1% congored solution and washed with 1M NaCl. The plates were incubated at 37°C for 24–48 h. Fifteen morphologically distinct bacterial isolates forming halo zones in CMC Congo red agar medium were considered and named as cellulose degrading bacteria, CDB1–15 (**Fig. 34**). C_x cellulase producing activities of the isolates were estimated by the carboxymethyl cellulose hydrolysis capacity (HC value) on the cellulose Congo red agar, i.e. ratio of diameter of clearing zone and colony. Bacteria showing high HC values were selected as cellulose degrading bacteria. The result showed that maximum Clearing zone size ranged between 0.9 to 2.1 cm where as HC value ranged between 1.25 to 2.5 respectively (**Table 35**) demonstrating that all the isolates have the ability to degrade the carboxymethyl cellulose and indicating high ability of Cx cellulase production (Table-3). Highest Cx cellulase production ability estimated by CMC hydrolysis capacity was observed with bacterial isolate CDB-5, (HC value 2.5) followed by CDB-12 (HC value 2.1) where as least HC value was observed with the bacterial isolate CDB-7.

Table 35: Cx cellulase production as estimated by HC value of the fifteen bacterial isolates from mangrove soil of Mahanadi delta, Odisha, India

Bacterial Isolates	Cellulase production ability	Colony size (in cm) (a)	Zone size (in cm) (b)	HC Value (Ratio of zone size and colony size) (b/a)
CDB-1	+	0.9	1.8	2.0
CDB -2	+	1.1	1.8	1.636
CDB -3	+	0.8	1.4	1.75
CDB-4	+	1.0	2.0	2.0
CDB-5	+	0.8	2.0	2.5
CDB-6	+	1.0	2.0	2.0
CDB-7	+	1.4	1.8	1.28
CDB-8	+	1.1	2.0	1.818
CDB-9	+	1.0	1.3	1.3
CDB-10	+	1.0	1.7	1.7
CDB-11	+	0.7	0.9	1.285
CDB-12	+	1.0	2.1	2.1
CDB-13	+	1.0	2.0	2.0
CDB-14	+	0.9	1.3	1.444
CDB-15	+	1.2	1.5	1.25

+ = Cellulase production ability

4.3.2. Extraction and estimation of crude cellulase enzyme

Growth medium was prepared in 250 ml Erlenmeyer flask containing 50ml of CMC broth medium. The flasks were sterilized at 121 °C for 15 min. The flasks were inoculated with 1% of standard inoculums (v/v) transferred to the production medium and then incubated on a rotary shaker (100 rpm). The supernatant of the culture medium was taken out at different time intervals by centrifugation at 10,000 rpm for 10 min at 4 °C. The broth after cultivation was used for crude enzyme studies (Dien *et al.*, 2006). The production of enzyme at different interval was determined to know the suitable incubation time for enzyme production.

The crude CMCase enzyme production activity was assayed by using method described by Samira *et al.* (2011). The enzyme activity was estimated by mixing 1.0 ml of enzyme solution with 1.0 ml of 1% (w/v) carboxymethyl cellulose (CMC) in 10mM sodium phosphate buffer, pH 7.0 as a substrate. The reaction was carried out at 37 °C for 60 min and stopped by adding 1.0 ml of 3,5-dinitrosalicylic acid (DNS) reagent. The reaction mixture was then boiled at 100 °C for 10 min and after cooling, OD was taken with spectrophotometer (systronics 119) at 546 nm. One unit of CMCase activity was expressed as 1 µ mol of glucose liberated per ml of enzyme per minute. The values obtained are compared with glucose standard curve.

Isolated 15 cellulose degrading bacterial isolates were further screened in broth medium supplemented with CMC cellulose to determine the cellulase activity. This is a most quantitative assay method used to determine the cellulase activity measured by determining the reducing sugar released using Dinitrosalicylic (DNS) method. Although congored test was sensitive enough for primary screening and isolation of cellulose degrading bacteria, but the clear zone width doesn't imply the amount of cellulase activity. Hence experiments were also carried out to determine the quantitative cellulase production ability of these 15 bacterial isolates (CDB-1, CDB-2, CDB-3, CDB-4, CDB-5, CDB-6, CDB-7, CDB-8, CDB-9, CDB-10, CDB-11, CDB-12, CDB-13, CDB-14 and CDB-15) up to 120 h of incubation (Table 36). The amount of cellulase activity by the bacteria in the culture broth was ranged between 2.471 and 98.253 U/ml. Cellulase activity of all the bacterial isolates showed an increasing trend up to a certain incubation period after which it was declined. Though all the isolates showed the cellulase activity however, CDB-12 showed highest cellulase activity followed by CDB-5 at 72 h of incubation (Fig. 51). Among these two bacterial isolates, CDB-12 showed maximum cellulase production ability of 98.253 U/ml followed by CDB-5 which showed cellulase production ability of 96.374 U/ml, after 72 h of incubation (Table 36). These two bacteria (CDB-5 and CDB12) were selected for optimisation of cellulase production.

Table 36: Screening of cellulase production activity by different bacterial isolates incubated upto 120 h of incubation

Bacterial Isolates	Cellulase production (U/ml) at different incubation period				
	24 h	48 h	72 h	96 h	120 h
CDB-1	2.471 ± 1.72	11.223 ± 1.92	17.298 ± 1.55	17.235 ± 1.63	17.321 ± 1.76
CDB-2	17.298 ± 1.61	17.707 ± 1.55	59.307 ± 1.10	38.53 ± 1.51	39.307 ± 1.76
CDB-3	13.707 ± 1.65	12.471 ± 1.68	22.24 ± 1.31	6.23 ± 1.58	2.471 ± 1.62
CDB-4	31.236 ± 1.64	58.701 ± 1.05	53.747 ± 1.71	36.471 ± 1.61	32.124 ± 1.62
CDB-5	21.235 ± 1.65	29.653 ± 1.82	96.374 ± 1.41	68.418 ± 1.54	54.942 ± 1.62
CDB-6	8.345 ± 1.71	11.126 ± 1.53	10.456 ± 1.39	9.844 ± 1.73	24.71 ± 1.41
CDB-7	6.112 ± 1.65	21.322 ± 0.98	28.942 ± 1.55	34.971 ± 0.85	34.595 ± 1.76
CDB-8	12.36 ± 1.52	44.48 ± 1.21	12.36 ± 1.63	9.884 ± 1.73	2.471 ± 1.79
CDB-9	8.321 ± 1.33	12.853 ± 1.43	18.53 ± 1.87	14.235 ± 1.36	11.43 ± 1.98
CDB-10	13.591 ± 1.42	13.591 ± 1.41	12.471 ± 1.61	12.432 ± 1.92	11.235 ± 1.53
CDB-11	16.18 ± 1.68	22.471 ± 2.13	32.124 ± 1.71	18.043 ± 1.66	7.413 ± 1.41
CDB-12	83.787 ± 1.81	88.48 ± 1.93	98.253 ± 1.38	97.787 ± 1.81	80.04 ± 1.92
CDB-13	14.942 ± 1.53	29.653 ± 1.15	24.942 ± 1.86	21.235 ± 1.54	22.24 ± 0.89
CDB-14	21.235 ± 2.1	63.013 ± 1.66	49.769 ± 1.85	21.235 ± 1.66	23.706 ± 1.83
CDB-15	2.471 ± 1.54	16.18 ± 1.43	27.298 ± 1.21	18.235 ± 1.51	17.298 ± 1.52

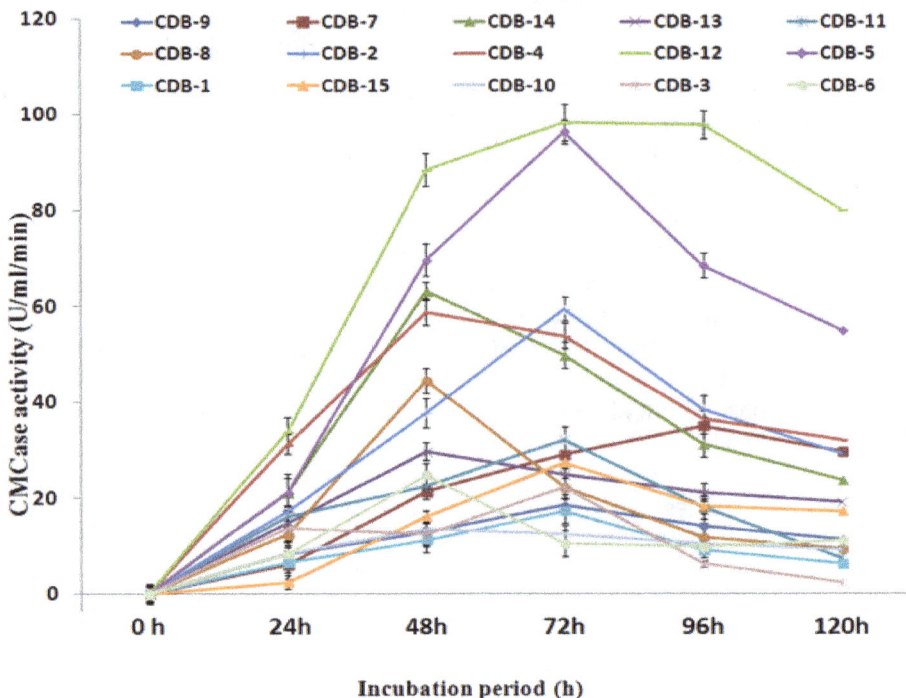

Figure 51. Cellulase production ability of the fifteen cellulose degrading bacterial isolates from mangrove soil of Mahanadi river delta, Odisha, India.

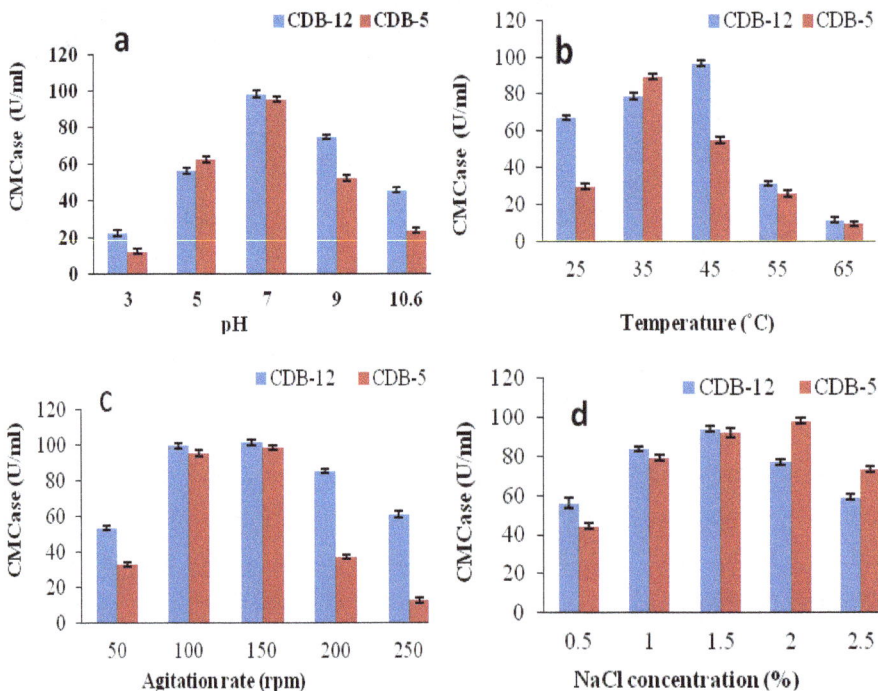

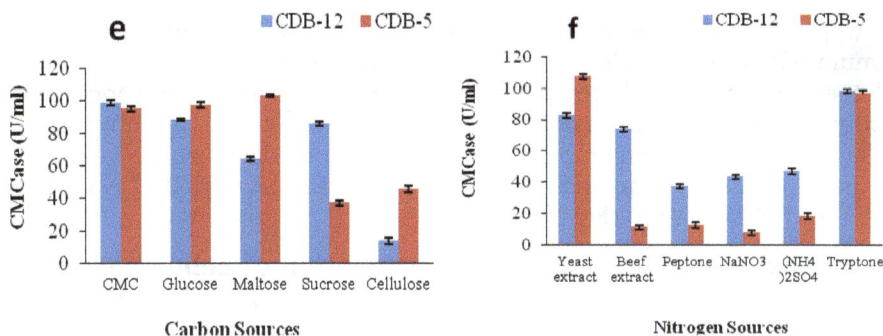

Fig. 52. Effect of (a) pH, (b) temperature, (c) agitation rate (rpm), (d) carbon sources, (e) nitrogen sources, and (f) NaCl concentration (%).

4.3.3. Optimization of parameters for production of cellulase

Optimization of cellulase production was carried out by inoculating the bacterial culture in the growth medium with respect to different environmental conditions such as in different pH, temperature, agitation rate, salt (NaCl) concentrations, carbon sources, and nitrogen sources and cellulase production were estimated following the standard method described by Samira *et al.* (2011). As the maximum cellulase production by both the isolates were observed after 72 h of incubation, the effect of different growth conditions such as pH, temperature, agitation rate along with carbon, nitrogen and salt (NaCl) sources was studied for optimum cellulase production in flask condition after 72 h of incubation. The effect of pH in the growth medium showed a gradual increase in cellulase production from 5.0 to 7.0 by both the strains. The cellulase production reached its maximum value at pH 7.0 by both the isolates, CDB-12 (98.253 U/ml) and CDB-5 (96.373 U/ml) (**Fig. 52a**).

For microbial growth and enzyme production, temperature is an important factor. In the present study, the bacterial isolate, CDB-12, produced maximum cellulase (99.538 U/ml) at 45°C (at 72 h of incubation and keeping the initial pH 7.0) whereas 35°C was found to be the optimum temperature for maximum cellulase production (96.373 U/ml) for the isolate CDB-5 (Fig. 52b). A significant decline in cellulase production was observed beyond the optimum temperature for both the organisms. The effect of agitation rate on the production of cellulase was optimized by keeping the culture condition up to 72 h of incubation, at pH 7.0 and at respective optimum temperature found for both the isolates. It was observed that both the isolates CDB-5 and CDB-12 showed maximum cellulase production of 99.47 U/ml and 101.253 U/ml, respectively, at 150 rpm (Fig. 52c). The effect of different carbon sources such as maltose, glucose, sucrose, and cellulose was supplied to the growth medium of two bacterial isolates separately by replacing the original carbon source CMC to optimize the cellulase production. It was observed that no such carbon sources were found suitable for maximum cellulase production than original carbon source CMC by both the isolates. However, in above-optimized condition and in the presence of CMC, both the bacterial isolates CDB-12 (101.753 U/ml/min) and CDB-5 (99.84 U/ml/min) showed maximum cellulase production (Fig. 52d). Different nitrogen sources were supplied under above optimized conditions in the broth medium to optimize the cellulase production. The isolate, CDB-12 showed highest enzyme production (102.253 U/ml/min) when the medium was supplemented with tryptone whereas yeast extract was found to be the best nitrogen source for optimum cellulase production (101.45 U/ml/min) by the isolate, CDB-5 (Fig. 52e). With the above-optimized growth conditions, the effects of various concentrations of sodium chloride in a range of 0.5–2.5% (w/v) were tested for optimum cellulase production. The optimum amount of enzyme production was observed in 2% of sodium chloride supplemented medium by both the isolates CDB-12 (94.21 U/ml) and CDB-5 (98.281 U/ml) (Fig. 52f).

4.3.4. Partial purification & characterization of cellulase

The protein of the culture filtrates was precipitated overnight with 70% $(NH_4)_2SO_4$. The resultant precipitate was collected by centrifugation at 10,000 rpm for 10 min. After centrifugation, the enzyme was dissolved in phosphate buffer and was kept in dialysis membrane, in a beaker containing phosphate buffer (pH 7.0) for overnight. Phosphate buffer was changed after every 6 h and the dialyzed sample was used for further characterization. To estimate the molecular weight of the partially purified enzyme, SDS-PAGE was done using 5% stacking gel and 10% resolving gel according to the method of Laemmli (1970). Quantification of protein content of crude and partially purified cellulase was done following the method of Lowry *et al.*, (1951), with Bovine serum albumin as a standard (Sigma, Mannheim, Germany).

For characterization, the partially purified enzyme was subjected to different parameters such as pH, temperature, and substrate concentrations and enzyme activity was determined following the standard method of Samira *et al.* (2011). Partial purification of crude cellulase of both the isolates was performed by ammonium sulphate precipitation followed by dialysis and SDS-PAGE gel electrophoresis. After partial purification, band sizes of partially purified proteins were found approximately 55 kDa (CDB-12) and 72 kDa (CDB-5) (Fig. 53d). Results showed that the partially purified cellulase from the bacterial isolate, CDB-5, exhibited a protein content of 8.5 mg/ml with a specific activity of 12.60 U/mg which corresponds to 2.25-fold purification and 33.34% yield (Table 37). Partially purified cellulase from the bacterial isolate, CDB-12, exhibited a protein content of 7.2 mg/ml with a specific activity of 16.42 U/mg which corresponds to 2.79-fold purification and 36.11% yields.

To study the enzyme activity, the partially purified enzymes were characterized in different experimental conditions such as different pHs (3.0–10.6), temperatures (25–65°C), and different substrate concentrations ranging from 0.25% to 1.5% (w/v). Both the bacterial isolates CDB-12 (118.27 U/ml) and CDB-5 (105.718 U/ml) showed their optimum cellulase activity at pH 9.0. (Fig. 53a). The effects of temperature on partially purified cellulase activity of the bacterial isolate, CDB-12, were found to be optimum at 45°C (109.964 U/ml) whereas the partially purified cellulase of the isolate, CDB-5, showed optimum activity at 35°C (102.88 U/ml) and declined thereafter (**Fig. 53b**) Substrate specificity of the partially purified enzyme was performed by assaying the activity of the partially purified enzyme against different concentrations of CMC ranging from 0.25% to 1.5% (w/v). It was observed that the partially purified enzyme of CDB-12 and CDB-5 showed a significantly optimum activity of 109.427 U/ml and 102.124 U/ml, respectively, at 1% (w/v) substrate concentration. A saturated cellulase enzyme activity was observed with an increase in substrate concentration (Fig. 53c).

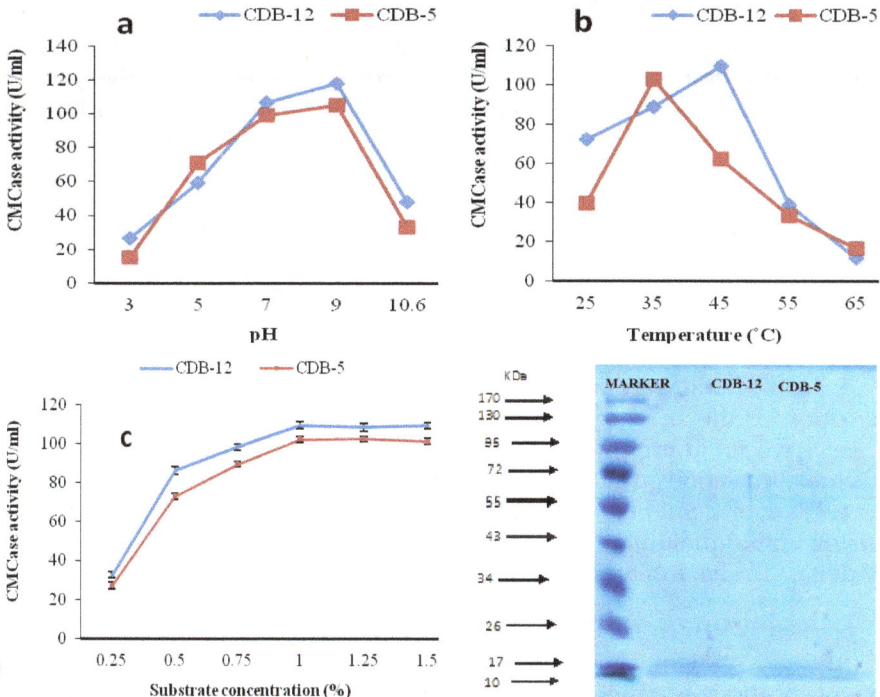

Fig. 53. Effect of (a) pH, (b) temperature, and (c) substrate concentration on partially purified cellulase activity. SDS-PAGE of partially purified cellulase from CDB-12 and CDB-5 (d).

Table 37 Partial purification of crude cellulase

Bacterial Isolates	Total Volume (ml)	Protein Content mg/ml	Total Protein (mg)	Cellulase activity (U/ml)	Total cellulase Activity (U)	Specific Activity (U/mg)	Fold of Puri-fication	Total Yield (%)
Culture extract of CDB-12	50.0	16.7	835	98.253 ± 1.88	4912.65	5.88	1.0	100%
70% (NH$_4$)$_2$SO$_4$ precipitation and dialysis of CDB-12	15.0	7.2	108	118.27 ± 1.47	1774.05	16.42	2.79	36.11%
Culture extract of CDB-5	50.0	17.0	850	96.374 ± 2.99	4818.7	5.66	1.0	100%
70% (NH$_4$)$_2$SO$_4$ precipitation and dialysis of CDB-5	15.0	8.5	127.5	107.124 ± 1.5	1606.86	12.60	2.25	33.34%

4.4. Sulphur Oxidation

The biochemical significance of sulphur is tremendous. Sulphur is required because of its structural role in the aminoacids cysteine and methionine and it is present in number of vitamins, such as thiamine, biotin and lipoic acid, as well as in coenzyme A (Madigan *et al.*, 2000). Mangrove soils are anaerobic environments rich in sulphate and organic matter. Mangrove soils are sulphidic and variable, since their chemistry is regulated by a variety of factors such as texture, tidal range and elevation, redox state, bioturbation intensity, forest type, temperature and rainfall (Alongi *et al.*, 1992). Different mangrove forest differs in their sulphide concentration because the roots (pneumatophores) of some species of mangrove can oxidise the surrounding soil while others do not have this ability (Nickerson and Thibodeauarticularly, 1985; Lacerda *et al.*, 1993). Sulphur oxidation improves soil fertility. It results in the formation of sulphate, which can be used by the plants, while the acidity produced by oxidation helps to solubilise plant nutrients and improves alkali soils. Although the sulphur cycle is one of the major factors in this ecosystem, little is known regarding the sulphur bacteria communities in mangrove soils (Varon-Lopez (2014). Although several studies have been made on microbial diversity of mangrove ecosystem, knowledge of sulphur oxidising bacterial communities in mangrove sediments is very sparse. A more thorough description of the sulphur oxidising bacterial diversity and distribution in a mangrove wduould improve our understanding of sulphur geochemistry as well as microbial metabolism of suphur in that ecosystem. Keeping the above in vision the present investigation is aimed to isolate, characterize and estimate the sulphur oxidising ability of sulphur oxidizing bacteria from mangrove soil of Mahanadi river delta, Odisha, India.

4.4.1. Isolation of sulphur oxidising bacteria

Sulphur oxidising bacteria were isolated from six different locations of mangrove soil sample of Mahanadi delta, Odisha using sulphur oxidizer-agar medium(g/l: 10 g of bacto-peptone, 1.5 g of K_2HPO_4, 0.75 g of ferric ammonium citrate and 1.0 g of $Na_2S_2O_3.5H_2O$). The initial pH was adjusted to 7.0 using 1 M HCl before sterilizing by autoclave. Agar was added to a final concentration of 15 g per liter. The plates were incubated at 30 °C for 24 h. Twenty eight morphologically distinct bacterial isolates forming distinct colony on agar medium were isolated and inoculated on thiosulphate broth medium containing bromophenol blue (BPB) as an indicator. Out of twenty eight isolates, 12 isolates were able to change the colour of the BPB in thiosulphate broth medium by reducing the pH of the medium (Fig. 54) and named as sulphur oxidising bacteria (SOB1-12). All the 12 bacterial isolates from the mangrove soil found to reduce the pH in thiosulphate broth medium. Out of the 12 bacterial isolates, two bacterial isolate SOB-7 (4.0) and SOB-8 (4.1) were found to decrease the pH of the medium more efficiently (Table 38).

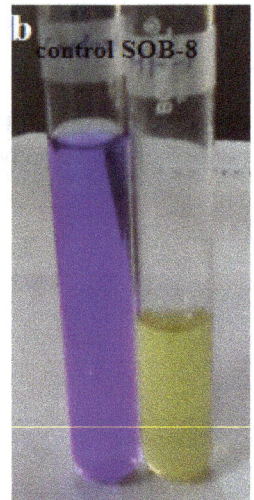

Fig.54: Growth of sulphur oxidising bacteria with control on thiosulphate broth supplied with bromophenol blue as an indicator after 3 days of inoculum

For qualitative screening of distinct sulphur oxidising bacteria, the isolated bacteria were further grown on the thiosulphate broth (Beijerinck, 1904) containing: 5.0 g $Na_2S_2O_3$, 0.1 g K_2HPO_4, 0.2 g $NaHCO_3$ and 0.1 g NH_4Cl in 1000 ml distilled water, with pH 8.0. Bromophenol blue was used as the indicator. The cultures which changed the colour of the thiosulphate broth from purple to colour less by reducing the pH after incubation for 3 days at 30°C were selected for further characterisation and evaluation of their sulphate ion and sulphide oxidase activity.

Table 38: Changes in pH of the thiosulphate bromophenol blue broth medium due to the growth and oxidation of sulphur by sulphur oxidising bacteria incubated up to 264 h of incubation

Bacterial Isolates	0 h	6 h	12 h	24 h	48 h	72 h	96 h	120 h	144 h	168 h	192 h	216 h	240 h	264 h
SOB-1	8.0	7.9 ± 0.13	7.4 ± 0.32	6.17 ± 0.77	5.75 ± 0.36	5.6 ± 0.43	5.47 ± 0.41	5.60 ± 0.39	5.50 ± 0.33	5.2 ± 0.33	5.4 ± 0.16	5.6 ± 0.27	5.8 ± 0.67	5.8 ± 0.36
SOB-2	8.0	7.8 ± 0.43	7.3 ± 0.79	7.66 ± 0.32	6.71 ± 0.24	6.72 ± 0.21	6.53 ± 0.13	6.3 ± 0.31	5.88 ± 0.73	5.29 ± 0.33	5.5 ± 0.19	5.56 ± 0.6	5.4 ± 0.37	5.7 ± 0.69
SOB-3	8.0	7.9 ± 0.23	7.6 ± 0.27	7.06 ± 0.33	6.81 ± 0.36	6.43 ± 0.16	5.85 ± 0.27	4.92 ± 0.36	4.89 ± 0.42	5.55 ± 0.32	5.6 ± 0.37	5.5 ± 0.23	5.4 ± 0.26	5.6 ± 0.37
SOB-4	8.0	6.7 ± 0.17	6.1 ± 0.56	6.2 ± 0.39	6.27 ± 0.21	5.83 ± 0.25	5.56 ± 0.41	5.59 ± 0.29	5.15 ± 0.14	5.2 ± 0.36	4.9 ± 0.87	4.7 ± 0.96	5.4 ± 0.53	5.3 ± 0.26
SOB-5	8.0	7.43 ± 0.5	7.08 ± 0.16	6.54 ± 0.19	6.31 ± 0.17	5.58 ± 0.15	5.48 ± 0.17	5.62 ± 0.38	5.1 ± 0.18	4.98 ± 0.26	4.9 ± 0.32	4.83 ± 0.37	5.42 ± 0.32	5.5 ± 0.46
SOB-6	8.0	7.4 ± 0.37	7.2 ± 0.23	6.43 ± 0.33	6.38 ± 0.25	5.83 ± 0.29	5.89 ± 0.49	5.39 ± 0.39	5.23 ± 0.26	5.61 ± 0.32	5.1 ± 0.32	4.71 ± 0.23	4.93 ± 0.37	4.97 ± 0.33
SOB-7	8.0	7.8 ± 0.23	7.5 ± 0.27	7.76 ± 0.37	7.32 ± 0.29	6.63 ± 0.38	6.19 ± 0.29	5.83 ± 0.18	5.88 ± 0.27	5.2 ± 0.26	4.1 ± 0.37	4.0 ± 0.27	4.1 ± 0.56	4.3 ± 0.37
SOB-8	8.0	7.8 ± 0.17	7.2 ± 0.16	6.88 ± 0.27	6.81 ± 0.38	5.96 ± 0.43	5.88 ± 0.33	5.34 ± 0.39	5.23 ± 0.29	4.35 ± 0.19	4.2 ± 0.33	4.1 ± 0.37	4.1 ± 0.47	4.6 ± 0.32
SOB-9	8.0	7.6 ± 0.26	7.5 ± 0.27	7.47 ± 0.18	6.67 ± 0.39	6.22 ± 0.19	6.20 ± 0.29	5.87 ± 0.49	5.45 ± 0.19	4.98 ± 0.34	4.7 ± 0.47	4.5 ± 0.47	4.76 ± 0.26	4.8 ± 0.32
SOB-10	8.0	7.8 ± 0.17	7.6 ± 0.13	7.22 ± 0.15	6.89 ± 0.59	611 ± 0.17	5.84 ± 0.37	5.79 ± 0.26	5.31 ± 0.29	5.1 ± 0.33	4.9 ± 0.32	4.8 ± 0.16	4.5 ± 0.27	4.6 ± 0.33
SOB-11	8.0	7.7 ± 0.36	7.7 ± 0.27	7.59 ± 0.28	7.15 ± 0.18	6.89 ± 0.19	6.42 ± 0.17	6.36 ± 0.27	5.79 ± 0.19	5.49 ± 0.26	4.4 ± 0.53	4.8 ± 0.26	5.15 ± 0.47	5.3 ± 0.27
SOB-12	8.0	7.5 ± 0.17	7.3 ± 0.22	7.0 ± 0.27	6.58 ± 0.13	6.42 ± 0.17	6.33 ± 0.34	5.79 ± 0.57	5.20 ± 0.23	4.9 ± 0.37	4.6 ± 0.13	4.7 ± 0.33	4.83 ± 0.36	4.9 ± 0.16

4.4.2. Sulphate ion determination

The amount of sulphate ion (SO_4^{2-}) produced during growth of sulphur-oxidizing bacteria on thiosulphate broth medium was determined spectrophotometrically. Sulphate was measured by adding 1:1 barium chloride solution (10% w/v) with supernatant followed by mixing the suspensions vigorously (Cha *et al.,*, 1999). A resulting, white turbidity due to barium sulphate formation was measured at 450 nm with spectrophotometer (Systronics 119). Potassium sulphate (K_2SO_4) was used as standard to construct a sulphate calibration curve (Kolmert *et al.,* 2000).

The amount of sulphate ion produced in the medium up to 264 h of incubation is presented in Table 39 and Fig. 55. Production of sulphate ion in the liquid medium by different strain was accompanied by significant drop in pH values from an initial pH of 8.0. The sulphate ion concentration in the medium was in between 25 mg/ml to 245 mg/ml with variation among different isolates at different incubation period. In the control no sulphate ion was detected as well no drop in pH was observed. However among the 12 bacterial isolates, the maximum sulphate ion production was observed by the bacterial isolate, SOB-7 (245 mg/ml) with maximum decrease in pH (pH 4.0) of the medium followed by the bacterial isolate, SOB-8 (240 mg/ml) with decrease in pH value of 4.1 of the thiosulphate broth medium.

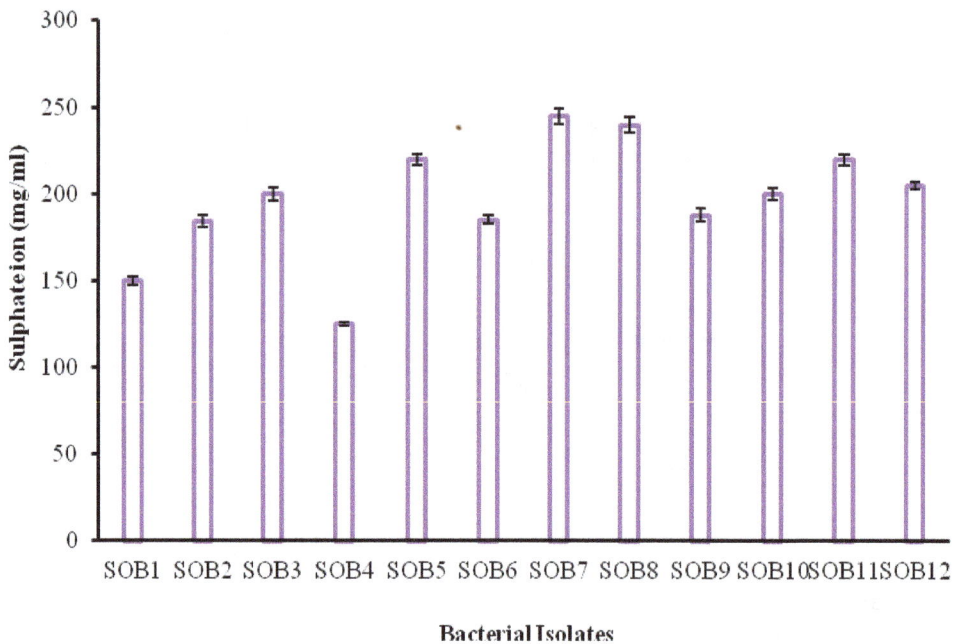

Fig.55: Quantitave evaluation of sulphate ion production ability (mg/ml) of twelve sulphur oxidising bacteria isolated from mangrove soil of Mahanadi delta.

Table 39: Quantitative estimation of sulphate ion (mg/ml) produced from twelve sulphur oxidising bacterial isolates

Bacterial isolate	24 h	48 h	72 h	96 h	120 h	144 h	168 h	192 h	216 h	240 h	264 h
SOB1	52 ± 2.0	60 ± 1.26	65 ± 2.98	115 ± 2.3	110 ± 2.4	138 ± 3.5	146 ± 1.2	150 ± 2.6	130 ± 2.2	75 ± 1.76	76 ± 1.13
SOB2	51 ± 1.33	50 ± 1.76	90 ± 2.76	97 ± 2.87	130 ± 2.6	142 ± 3.3	176 ± 1.2	184 ± 3.5	129 ± 2.3	105 ± 1.6	115 ± 2.8
SOB3	93 ± 2.66	152 ± 2.33	173 ± 1.32	185 ± 1.7	180 ± 2.4	184 ± 1.6	185 ± 2.4	200 ± 3.7	150 ± 2.9	160 ± 1.5	158 ± 2.5
SOB4	35 ± 3.33	46 ± 1.78	43 ± 1.33	54 ± 1.65	66 ± 2.87	85 ± 1.74	85 ± 2.52	120 ± 1.2	125 ± 2.3	74 ± 1.64	78 ± 1.36
SOB5	80 ± 1.54	85 ± 1.95	145 ± 1.43	154 ± 1.8	173 ± 2.3	175 ± 1.5	180 ± 2.6	185 ± 1.3	220 ± 3.2	205 ± 1.4	177 ± 1.5
SOB6	45 ± 1.67	39 ± 2.43	55 ± 1.55	85 ± 1.99	94 ± 1.43	115 ± 1.8	160 ± 1.8	185 ± 2.4	140 ± 1.6	130 ± 2.2	130 ± 1.2
SOB7	30 ± 1.76	65 ± 2.33	90 ± 1.33	140 ± 1.7	165 ± 1.0	169 ± 1.6	187 ± 1.3	205 ± 1.4	245 ± 4.3	240 ± 2.9	220 ± 1.5
SOB8	25 ± 1.43	40 ± 2.32	98 ± 1.76	120 ± 1.9	171 ± 2.3	183 ± 2.7	195 ± 2.2	225 ± 1.2	227 ± 1.2	240 ± 4.4	200 ± 1.2
SOB9	65 ± 1.54	67 ± 2.54	70 ± 1.19	70 ± 1.98	85 ± 2.55	89 ± 2.92	121 ± 2.1	170 ± 2.7	188 ± 3.7	170 ± 1.5	130 ± 1.5
SOB10	44 ± 1.34	50 ± 2.21	70 ± 1.33	77 ± 1.53	69 ± 1.93	75 ± 1.38	85 ± 1.76	140 ± 2.1	185 ± 1.3	200 ± 3.4	175 ± 1.9
SOB11	40 ± 1.43	48 ± 2.43	54 ± 1.32	86 ± 1.98	105 ± 1.7	130 ± 1.4	185 ± 1.8	220 ± 3.3	208 ± 1.2	175 ± 1.9	165 ± 1.9
SOB12	70 ± 1.75	122 ± 2.33	135 ± 1.39	149 ± 1.99	180 ± 2.11	180 ± 1.38	200 ± 1.89	205 ± 2.39	175 ± 1.39	170 ± 1.57	160 ± 1.0

4.4.3. Sulphide oxidase assay

The sulphide oxidase activity was determined by measuring the product of enzymatic reaction, sulphate (SO_4^{2-}) in the reaction mixture following the standard method of Hirano et al., (1996). The reaction was initiated by addition of 0.5 ml of sodium sulphide (Na_2S) solution into the reaction mixture that contain 4.5 ml of 0.1 M sodium acetate buffer (pH 5.6) and 1 ml supernatant. The Na_2S solution was prepared by dissolving 0.06 g Na_2S in an alkaline solution consisting of 0.16 g NaOH, 0.02 g EDTA $Na_2.2H_2O$ (sodium ethylene diamine tetra acetic acid), 2 ml glycerol and 40 ml distilled water. The Na_2S solution was freshly prepared prior to use. The mixture was incubated for 30 minutes at 30°C and the reaction was subsequently terminated by the addition of 1.5 ml NaOH (1.0 M) followed by thorough mixing. Concentration of sulphate ion formed during sulphide oxidase assay was detected by the reaction of equal volume of barium chloride solution (10% w/v) and reactant and the absorbance was measured at 450 nm using spectrophotometer. The measurement of sulphate ion in the sample was based upon the formation of barium sulphate after addition of barium chloride which leads to the white turbidity. The amount of turbidity formed is proportional to the sulphate ion concentration in the sample. One unit of sulphide oxidase activity was defined as amount of enzyme required to produce 1 μmol sulphate per hour per ml (U/ml).

Experiments were carried out to determine the sulphide oxidase production activity (S.O. activity) of 12 bacterial isolates (SOB-1, SOB-2, SOB-3, SOB-4, SOB-5, SOB-6, SOB-7, SOB-8, SOB-9, SOB-10, SOB-11 and SOB-12) during 264 hours of incubation (**Table 40**). Their sulphide oxidase production efficiency ranged from 11.6 U/ ml to 126.83 U/ml (**Table 40**). All the isolates showed the sulphide oxidase production activity however, SOB-7 and SOB-8 showed comparatively better sulphide oxidase production activity than others (**Fig.56**). Among these twelve isolates, SOB-8 showed comparatively highest sulphide oxidase production activity (126.83 U/ml) followed by the isolate, SOB-7 (126 U/ml) (**Fig. 56**).

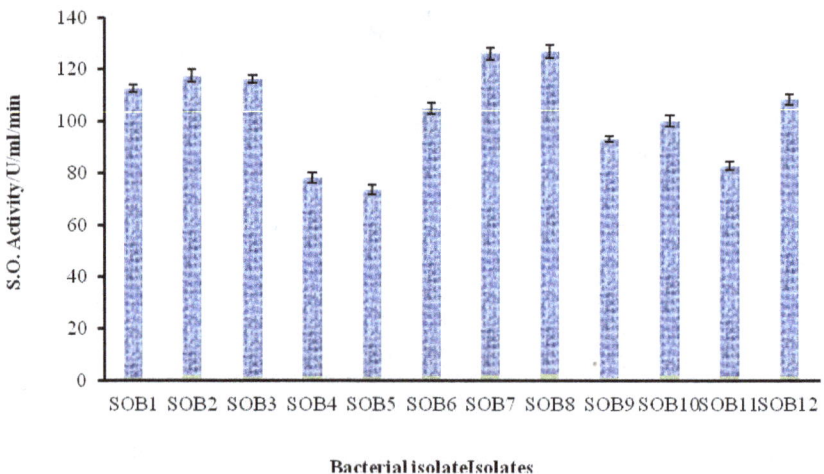

Fig. 56: Sulphide oxidase activity of different bacteria isolates from mangrove soil of Mahanadi river delta, Odish

Table 40: Quantitative screening of sulphide oxidase (U/ml) production by different bacterial isolates.

Bacterial Isolates	24 h	48 h	72 h	96 h	120 h	144 h	168 h	192 h	216 h	240 h	264 h
SOB1	26.28±1.6	40.96±1.7	53.94±1.4	65.41±1.8	67.34±0.9	69.4±1.53	70.4±0.92	82.34±1.4	112.6±1.3	112.6±1.4	106.77 ± 1.8
SOB2	28.16±1.2	37.5±1.63	65.7±1.85	66.88±1.9	76.26±2.3	77.74±1.7	83.3±0.85	86.82±1.3	106.7±1.1	116.1±2.2	117.5 ± 2.29
SOB3	38.72±0.8	52.8±1.43	58.66±2.2	69.4±1.9	75.09±1.8	116.1±1.4	97.38±2.3	79.78±1.2	82.13±1.6	83.3±1.32	68.05±1.8
SOB4	22.29±1.6	23.33±1.2	32.66±1.7	32.66±1.7	33.83±1.7	33.83±1.9	44.33±1.2	50.16±1.3	53.66±1.7	78.16±1.9	77.00±1.66
SOB5	31.5±1.54	46.66±1.3	47.83±1.3	51.33±1.8	52.5±1.73	54.83±1.6	64.16±1.6	66.5±1.38	65.33±1.7	73.5±1.73	74.66±1.8
SOB6	11.66 ± 1.3	45.5±1.28	61.8±1.2	56±1.72	56±1.99	58.33±1.6	82.83±1.3	84±1.39	96.83±2.3	105±2.1	103.83 ± 2.3
SOB7	40.83±1.8	53.66 ± 1.9	55.33±1.6	61±1.94	66±1.53	71.83±1.3	85.33±1.8	99.4±1.36	115.3±1.6	126±2.26	119.00 ± 2.33
SOB8	30.33±0.6	61.5±1.08	78±1.95	81±1.36	93.33±1.5	105.3±1.5	111.1±1.6	121.6±1.6	126.8±2.6	122.5±2.5	113.05 ± 0.87
SOB9	29.16±0.8	35±1.91	44.33±1.7	53.66±1.9	57.16±1.8	60.66±1.3	64.16±1.9	68.05±1.9	78.16±1.4	93.3±1.09	88.66 ± 1.65
SOB10	24.5±1.29	16.33±1.2	52.5±1.54	57.16±1.5	61.83±1.9	61.83±1.9	58.33±1.5	66.88±1.5	76.26±1.6	100.3±2.2	98.00 ± 0.99
SOB11	21±1.76	30.33±1.3	35±1.54	32.66±1.5	37.33±1.5	44.33±1.8	65.33±1.4	57.16±1.4	82.83±1.7	82.8±1.43	66.88±1.5
SOB12	11.6±1.1	43.16±1.8	65.33±1.7	66.88±1.9	68.83±1.8	96.83±1.6	88.66±1.5	78.16±1.3	98±1.83	108.5±1.9	108.50 ± 1.54

4.4.4. Selection of most efficient sulphur oxidising bacteria

Based on their ability to decrease the pH of the medium, maximum sulphate ion production and subsequent enzyme production capacity of all the 12 bacterial isolates, two bacterial isolates namely SOB-7 (**Fig. 57**) and SOB-8 (**Fig.58**) were selected as most efficient sulphur oxidising bacteria. A comparative analysis of sulphate ion and sulphide oxidase production is given in (**Fig. 57**) and (**Fig.58**). It was observed that the bacterial isolate SOB-7 produced maximum sulphate ion of 245 mg/ml which was associated with maximum sulphide oxidase production (126 U/ml) and decrease in pH of 4.0. Similarly, the SOB-8 shared maximum sulphate ion and sulphide oxidase of 240 mg/ ml and 126.8 U/ml, respectively associated with decrease in pH value of 4.1.

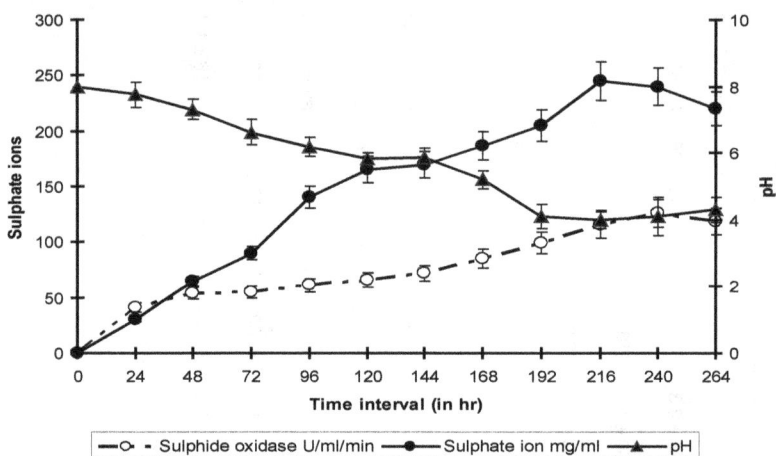

Fig. 57: Comparative pH reduction, sulphate ion production and sulphide oxidase production ability of bacterial isolate, SOB-7.

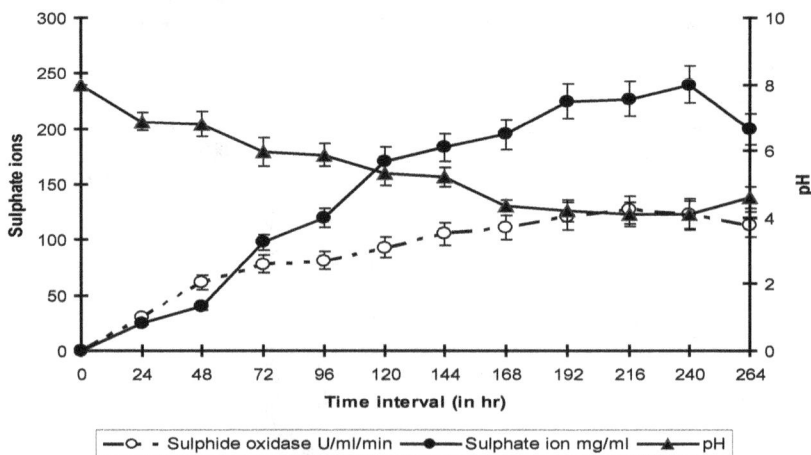

Fig. 58: Comparative pH reduction, sulphate ion production and sulphide oxidase production ability of SOB-8.

4.4.5. Optimisation of parameters for the production of sulphide oxidase

Optimisation of sulphide oxidase production were carried out by inoculating the bacterial culture in the growth medium with respect to different environmental condition such as pH, temperature, thiosulphate concentration, nitrogen source and enzyme activities were measured following the standard method of Hirano *et al.* (1996).

The sulphide oxidase production by these two selected isolates, SOB-7 and SOB-8 were carried out in different parameters such as pH (pH 3.0- 9.0), incubation temperature (25°C - 65°C), nitrogen sources (yeast extract, casein, ammonium chloride, and potassium nitrate) and different concentration of thiosulphate (5 – 25 mg/ml (w/v).

Effect of pH on enzyme production showed that very less sulphide oxidase produced by both the isolates when the initial pH of the culture medium was maintained at pH 3.0 (**Fig. 59a**). Maximum sulphide oxidase production (123.66 U/ ml) was observed by the isolate SOB-7 when the initial pH of the culture medium was maintained at pH 7.0. In case of the isolate SOB-8, sulphide oxidase production was found to be maximum (121.33U/ml) when the initial pH of the medium was maintained at pH 9.0. Decrease in enzyme activity was observed beyond their optimum pH.

Optimum temperature for the production of sulphide oxidase by both the strains was observed at 45°C (**Fig. 59b**). The bacterial isolate, SOB-7 showed maximum sulphide oxidase activity of 125.0 U/ml, where as maximum sulphide oxidase activity of 120 U/ml was observed at temperature of 45°C by the isolate, SOB-8. Sulphide oxidase productions decreased by both the bacterial isolates with increase in temperature beyond 45°C.

Effect of nitrogen source towards sulphide oxidase synthesis was studied by amended sulphur-oxidizer medium with varying nitrogen sources (**Fig.59c**). The initial supplementation of culture medium with peptone showed enhanced sulphide oxidase production (126.83 U/ml) by SOB-8 and (126.0 U/ml) by SOB-7 in comparison to other source used.

Concentration of thiosulphate in sulphur-oxidizer medium was varied in a range of 5– 25 mg/ml (w/v). Sulphide oxidase production was significantly affected by the initial increased in thiosulphate concentration. Maximum sulphide oxidase productions of 125.1 U/ml by SOB-8 and 137.32 U/ml by SOB-7 were observed when 10 mg/ml of thiosulphate was incorporated in the medium (**Fig. 59d**).

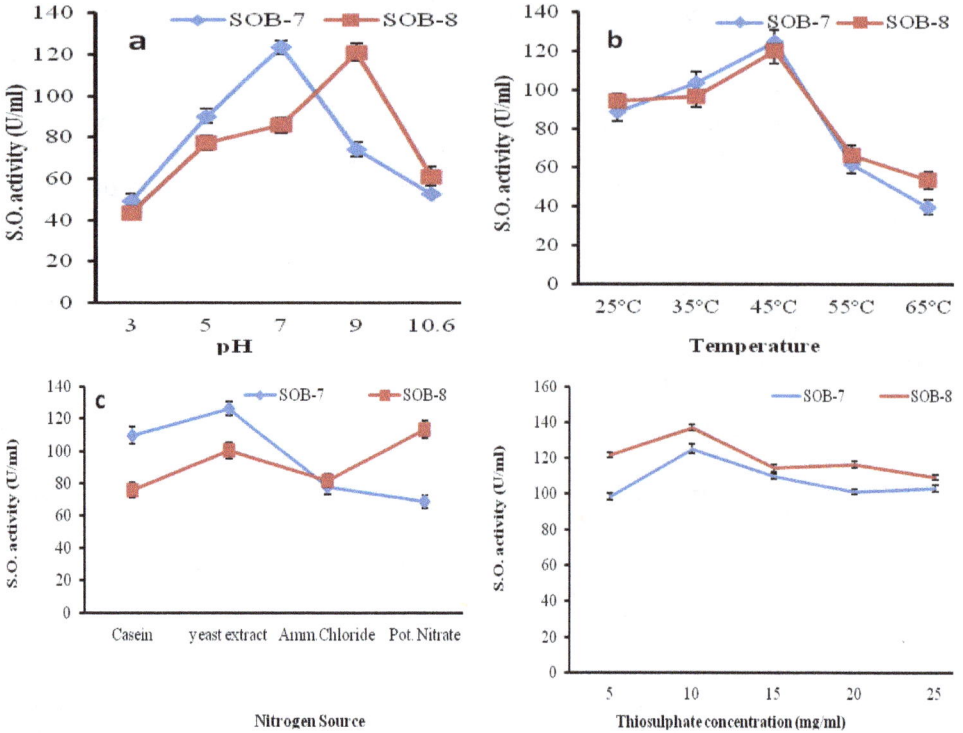

Fig 59: Effect of (a) pH, (b) Temperature, (c) Nitrogen source, (d) Thiosulphate concentration on sulphide oxidase production activity

4.4.6. Purification of sulphide oxidase & Molecular weight determination

Partial purification of sulphide oxidase was carried out by ammonium sulphate precipitation followed by dialysis. The gradient ammonium sulphate precipitation (30 to 85%) was carried out with chilled cell free culture broth. The precipitate was collected by centrifugation after two hours incubation at 4 °C and dissolved in 0.2 M phosphate buffer (pH 7). The enzyme extract was dialyzed overnight against the same buffer at 4 °C and the dialyzed enzyme was used for further studies. The protein concentration of both the isolates was estimated by Bradford method (1976). Electrophoretic analysis of extracellular sulphide oxidase from both the isolates has been carried out.

Results are presented in **Table 41.** Sulphide oxidase from the isolate, SOB-7 could be purified 2.89 fold with 36.74% yield and specific activity of 16.77 U/mg. Similarly Sulphide oxidase from the bacterial isolate, SOB-8 was purified 2.38 fold with 33.11% yield and specific activity of 11.2 U/mg.

Table 41: Partial purification of sulphide oxidase

Isolates	Total Volume (ml)	Protein mg/ml	Total Protein (mg)	Sulphide oxidase U/ml	Total Activity (U)	Specific Activity U/ mg	Fold of Purification	Total Yield (%)
Culture extract of SOB-7	50	21.7	1085	126 ± 2.26	6300	5.80	1	100
70% $(NH_4)_2SO_4$ precipitation and dialysis of SOB-7	15	9.2	138	154.31 ± 2.3	2314.6	16.77	2.89	36.74
Culture extract of SOB-8	50	27	1350	126.83 ± 2.6	6341.5	4.69	1	100
70% $(NH_4)_2SO_4$ precipitation and dialysis of SOB-8	15	12.5	187.5	140 ± 6.32	2100	11.2	2.38	33.11

Crude enzyme SDS-PAGE results of both the isolates showed the presence of multiple bands since along with sulphide oxidase, some other proteins can be produced by the organisms. But after partial purification the enzyme of the isolate SOB-7 showed some specific bands of approximately 72 kDa, 55 kDa, 50 kDa and 43 kDa. Partially purified protein of SOB-8 revealed some specific band of approximately 30 KDa and 25 kDa.(**Fig. 60**).

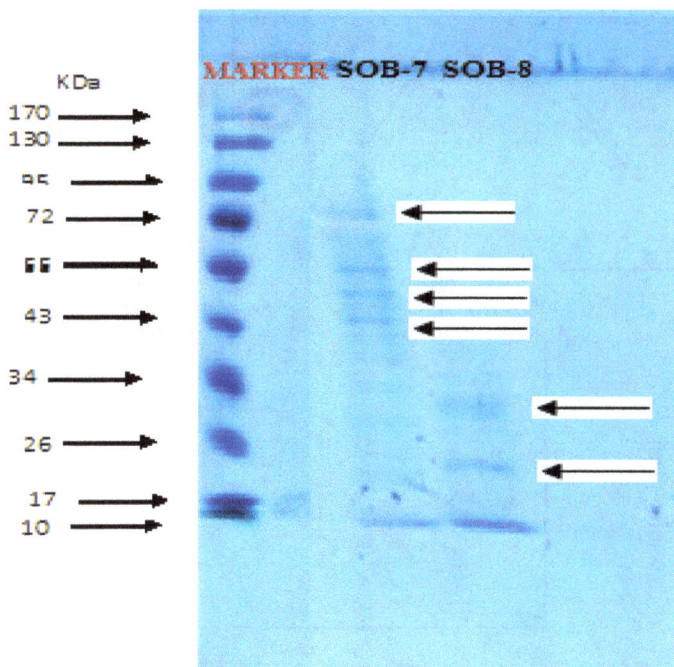

Fig. 60: Partially purified sulphide oxidase profile on SDS PAGE.

4.4.7. Characterisation of partially purified sulphide oxidase

For characterization, the partially purified enzyme was subjected to different parameters such as pH, temperature and substrate concentration and sulphide oxidase activity was measured following the standard method of Hirano *et al.* (1996)

The effects of pH on sulphide oxidase activity of the partially purified enzyme produced by the isolates were examined at various pH ranging from pH 3.0 to 10.6 in different buffer solution. The optimal pH for sulphide oxidase activity of the isolate, SOB-7 was found at pH 7.0 (136.66U/ml), whereas pH 9.0 was found optimum (131.88U/ml) for sulphide oxidase activity of SOB-8 (**Fig. 61a**).

The effects of temperature on partially purified sulphide oxidase activity of both the isolates were examined at various temperatures ranging from 25°C to 65°C. Optimal temperature for partially purified sulphide oxidase activity of both the isolate was found to be 45°C, beyond which steady decrease in enzyme activity was observed (**Fig. 61b**).

Substrate specificity of the purified enzyme was performed by assaying the activity of the purified enzyme against different concentrations of Na$_2$S solution (0.25-2.5 mg/ml) to the reaction mixture. The enzyme activity of both the isolates increased exponentially up to a certain point but remains almost same thereafter. It was observed that the partially purified enzyme of the both the isolate, SOB-7 showed significantly higher activity at 1.5 mg/ml of substrate concentration (**Fig. 61c**).

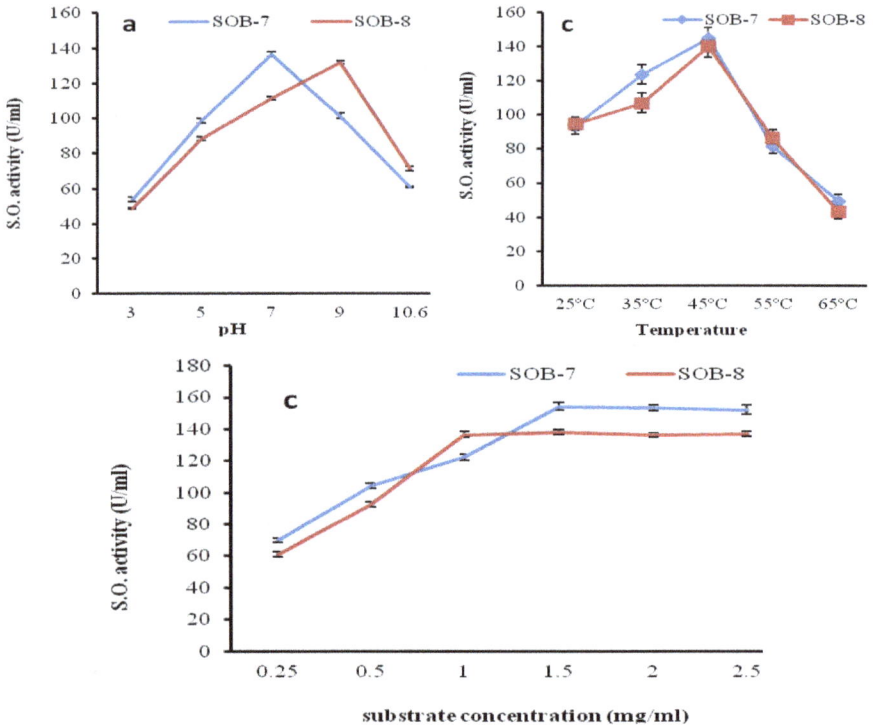

Fig.61: Effect of (a) pH, (b) temperature, (c) substrate concentration on partially purified sulphide oxidase activity.

4.5. Evaluation of Enzyme Production from Halotolerant Bacterial Strains

Marine microorganisms which are salt tolerant, provide an interesting alternative for therapeutic purposes. Marine microorganisms have a diverse range of enzymatic activity and are capable of catalyzing various biochemical reactions with novel enzymes. Especially, halophilic microorganisms possess many hydrolytic enzymes and are capable of functioning under conditions that lead to precipitation of denaturation of most proteins (Ventosa and Nieto, 1995). Further, it is believed that sea water, which is saline in nature and chemically closer to the human blood plasma, could provide microbial products, in particular the enzymes, that could be safer having no or less toxicity of side effects when used for therapeutic applications to humans (Sabu, 2003). A number of extra- and intracellular enzymes from moderately halophilic bacteria have been isolated and characterized. These include hydrolases (amylases, nucleases, phosphatases, and proteases), which are currently of commercial interest (Ventosa and Nieto, 1995).

4.5.1. Extraction and assay of enzyme activity

The enzymes were extracted from the bacteria according to Selander *et al.* (1986). The bacterial pellet was washed three times by centrifugation at 5,000 g for 20 min with 8.5% (w/v) NaCl and finally with Tris-EDTA buffer (10 mM Tris containing 1 mM EDTA, pH 6.8). The final pellet was stored for 12 h at 20°C, macerated with sterilized glass beads or wool in 4 ml Tris-EDTA buffer on a chilled mortar and pestle placed on an ice bath, centrifuged (15,000 g, 4 ± 0.1°C, 15 min) and the enzyme activity of the supernatant was assayed. Heat killed enzyme (98 ± 1°C, 5 min) was used for all control sets of assay and the enzyme protein was estimated as BSA equivalent with coomassie reagent (Bradford, 1976). Catalase (CAT) activity was measured following decrease in absorbance for H_2O_2 at 230 nm (Kar and Mishra, 1976) and the units (U) (mol H_2O_2 decomposed/mg protein/min) of activity were calculated considering 23.04/mM/cm for H_2O_2 at 230 nm. The peroxidase (PO) and polyphenol oxidase (PPO) activities were determined from the increase in absorbance at 430 nm (Kar and Mishra, 1976) and the enzyme activities were expressed as units (μmol purpurogallin formed/mg protein/min) using Σ2.47/mM/cm at 230 nm for purpurogallin. The ascorbate peroxidase (AP) activity was determined from the decrease of absorbance at 290 nm and the activity units (mol ascorbate decreased/mg protein/min) were assayed considering Σ2.8/mM/cm for ascorbate at 290 nm (Nakano and Asada, 1981). The ascorbic acid oxidase (AAO) activity was recorded following decrease in absorbance at 265 nm (Mahadevan and Sridhar, 1986) and the activity units (M ascorbic acid decreased /mg protein/ min) were calculated from 14 mM/cm for ascorbic acid at 265 nm. All experiments were repeated three times.

The activities (U/mg protein/min) of the different oxidative enzymes of the organisms; catalase (CAT) (0.05 -0.41) (Fig. 62a), peroxidase (PO) (0.06 - 0.98) (Fig. 62b), polyphenol oxidase (PPO) (0.25 - 1.13) (Fig. 62c),ascorbate peroxidase (APO) (8.10 - 29.35) (BSB 2 has no activity) (Fig. 62d), and ascorbic acid oxidase (AAO) (0.02 - 39.43) (Fig. 62e) were variable and no correlation could be obtained with their level of stress tolerance. Among the enzymes, the AAO and APO activity was found more in BSB 1 (Fig. 60).

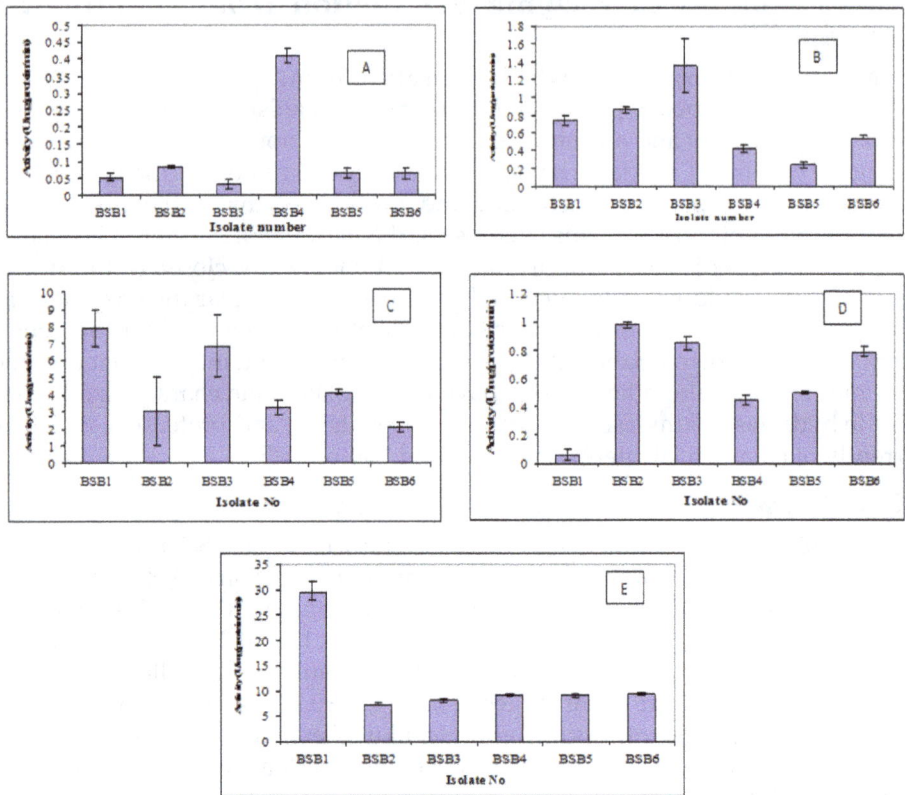

Fig62. Activity of (A)Catalase, (B) Peroxidase (C) Polyphenol oxidase (d) Ascorbate peroxidase and (e) Ascorbic acid oxidase enzymes from mangrove halotolerant bacterial isolates

4.6. Bioremediation of Heavy Metals

Many of these microbes possess unique capability to tolerate the hyper saline condition as well as various heavy metals and metalloids. Heavy metals are increasingly found in microbial habitats due to natural and industrial processes, for which microbes have evolved several mechanisms to tolerate the presence of heavy metals. Due to their high stress tolerance capacity these micro organisms are very useful for biotechnological applications in terms of bioremediation and biomineralization. The adaptation to heavy metal rich environments is resulting in microorganisms which show activities for biosorption, bioprecipitation, extracellular sequestration, transport mechanisms, and/or chelation. Such resistance mechanisms are the basis for the use of microorganisms in bioremediation approaches (Haferburg and Kothe, 2007). Contamination of the environment by heavy metals is a consequence of technological and industrial processes (Nriagu, 1996). This has led to the increasing concern about the effects of toxic metals as environmental contaminations.

Among the several bacterial species identified from mangrove ecosystem of Bhitarkanika, Odisha, finally three bacteria (two moderately halotolerant *Bacillus megaterium* (BSB6, BSB12) and one halophilic *Vigribacillus* sp. (HPB4)) were selected and evaluated for their biotechnological potentials. Thus all the three strains were screened for their ability to produce industrially important enzymes viz. amylase, cellulase, xylanase, keratinase, chitinase, protease and tannase. Besides their enzyme production, two halotolerant *B. megaterium* and one *Vigribacillus* sp. were screened for their heavy metal tolerance potential. Based on their heavy metal tolerance potential while the two *B. megaterium* were evaluated for selenite reduction ability the *Vigribacillus* sp. was evaluated for chromium reduction ability.

4.6.1. Heavy metal tolerance potential

Two halotolerant *Bacillus megaterim* (BSB6 and BSB12) and one halophilic *Vigribacillus* sp. isolated from mangrove soil of Bhitarkanika were screened for their heavy metal tolerance potential against seven heavy metal salts viz. $ZnSO_4$, $CuSO_4$, $MnSO_4$, $CdSO_4$, $CdCl_2$, Na_2SeO_3 and $K_2Cr_2O_7$. Two *B. megaterium* strains tolerated upto 0.75 mM $ZnSO_4$, 0.5 mM $CdSO_4$, $CdCl_2$ and 800 mM $CuSO_4$, but it was interesting to note that both the bacterial species could tolerate a very high concentration of selenite (1000 mM) in nutrient broth (**Table 42**). On the otherhand, the halophilic *Vigribacillus* sp. could able to tolerate up to 0.45 mM $ZnSO_4$, 850 mM $CuSO_4$, 800 mM $MnSO_4$, 0.75 mM $CdSO_4$, 075 mM $CdCl_2$, 900 mM Na_2SeO_3 and 1000 mM $K_2Cr_2O_7$. Due to high tolerance towards Na_2SeO_3 by *B. megaterium* and $K_2Cr_2O_7$ by *Vigribacillus* sp. these strains were further evaluated for their reduction kinetics.

Table 42: Screening of heavy metals (MIC) of selected bacterial species

Isolate No.	MIC (mM)						
	$ZnSO_4$	$CuSO_4$	$MnSO_4$	$CdSO_4$	$CdCl_2$	Na_2SeO_3	$K_2Cr_2O_7$
BSB6	750	800	750	500	500	**1200**	800
BSB12	750	800	750	500	500	**1200**	800
HPB4	450	850	800	750	750	900	**1000**

4.6.2. Reduction of selenium

Selenium is of considerable environmental importance as it is essential at low concentrations but toxic at high concentrations for animals and humans, with a relatively small difference between these values (Fordyce, 2005). Selenium occurs in different oxidation states as reduced form (selenide, Se^{2-}), least mobile elemental selenium (Se0) and water soluble selenite (SeO_3^{2-})/selenite (SeO_4^{2-}) oxyanions. The seleniferous agriculture drainage and effluents from thermal power stations, oil refineries, smelting plants, glass production, pigments and semiconductor industries are the major sources of water soluble selenium species in aquatic environment (Kashiwa *et al.*, 2000; Siddique *et al.*, 2006). As such it is necessary to develop suitable methods to reduce the concentrations of these oxyanions in waste streams to permissible limits before their discharge into aquatic environment. In addition to physicochemical methods (Zhang *et al.*, 2005; Rovira *et al.*, 2008), like chemical precipitation, catalytic reduction and adsorption/ ion exchange, there are several reports describing the bacterial reduction of selenite/selenate to less toxic elemental selenium and can form a viable and cost effective approach for abatement of excess selenium in contaminated water.Two Gram (+)^{ve} bacterial strains, BSB6 and BSB12, showing resistance and potential for Se(IV) reduction among 26 moderately halotolerant isolates from the Bhitarkanika mangrove soil were characterized by biochemical and 16S rDNA sequence analyses. Both of them were strictly aerobic and able to grow in a wide range of pH (4–11), temperature (4–40 ^{0}C) and salt concentration (4–12%) having an optimum growth at 37 ^{0}C, pH 7.5 and 7% salt (NaCl). The biochemical characteristics and 16S rDNA sequence analysis of BSB6 and BSB12 showed the closest phylogenetic similarity with the species *Bacillus megaterium*. Both the strains effectively reduced Se (IV) and complete reduction of selenite (up to 0.25 mM) was achieved within 40 h. SEM with energy dispersive X-ray and TEM analyses revealed the formation of nano size spherical selenium particles in and around the bacterial cells which were also supported by the confocal micrograph study. The UV–Vis diffuse reflectance spectra and XRD of selenium precipitates revealed that the selenium particles are in the nanometric range and crystalline in nature. These bacterial strains may be exploited further for bioremediation process of Se (IV) at relatively high salt concentrations and green synthesis of selenium nanoparticles.

The potential of BSB6 and BSB12 for Se (IV) reduction was studied by varying the concentration in the range 0.05–2.0 mM under optimized growth conditions (37 ^{0}C) and pH 7.5) (**Fig.63a**) The assay for selenium reduction was carried out in 100 mLNB as a static culture taken in 250 mL Erlenmeyer flasks and incubated at 37^{0}C in a BOD incubator. The BSB6 and BSB12 strains were cultured in the NB medium (pH 7.5), at 37 ^{0}C for 8 h, centrifuged (8000g, 10 min at 4 ^{0}C) and the pellet was resuspended in 2 mL of Tris–HCl buffer (pH, 8.0). The cell suspension (100 lL) was inoculated into the test medium (100 mL) containing 0.05–2.0 mM of SeO^{3}2. At a fixed selenite concentration (2.0 mM), the pH and salt concentration were also varied in the ranges 4–10 and 4–12% w/v, respectively. Control experiments without selenite were performed simultaneously. At regular intervals, 2 mL culture was withdrawn from each flask and centrifuged at 10,000g for 10 min at 10 ^{0}C and the remaining selenite content in the supernatant was analyzed. A representative time course of Se (IV) reduction along with control is shown in **Fig. 63a**. There is practically no change of selenite concentration in the control during the incubation period. However, both the bacterial strains effectively reduced Se(IV) to Se(0) which is evident from the change in colour of the reduction medium to orange–red/red due to generation of allotropic form of elemental selenium during the exponential growth phase. The growth kinetics under identical conditions of 0.25 mM initial concentration of selenite also indicated a lag phase of 3–4 h followed by a 4–15 h exponential phase and finally an unsteady stationary phase. Although the overall amount of Se(IV) reduction increased with an increase of Se(IV) concentration at a particular incubation period, almost complete reduction of Se(IV) was observed only at lower Se(IV) concentrations (0.05–0.25 mM). Analyses of precipitated fraction of the culture in these cases also corroborated the results of nearly complete transformation of selenite to elemental selenium. However, formation of a very small amount of selenide (Se^{2-}) during the reduction process cannot be ruled out. At higher Se(IV) concentrations the reduction rate decreases and is found incomplete within the studied incubation period presumably due to a lack of requisite cell growth for Se(IV) reduction. The other factor for lowering the rate of Se(IV) reduction at higher concentrations may be due to interaction of selenite with some organic components such as bacterial cell enzymes that inhibits Se(IV) reduction . It has been also observed that maximum reduction of Se(IV) reduction at 4% of Nacl **(Fig. 63b)**.

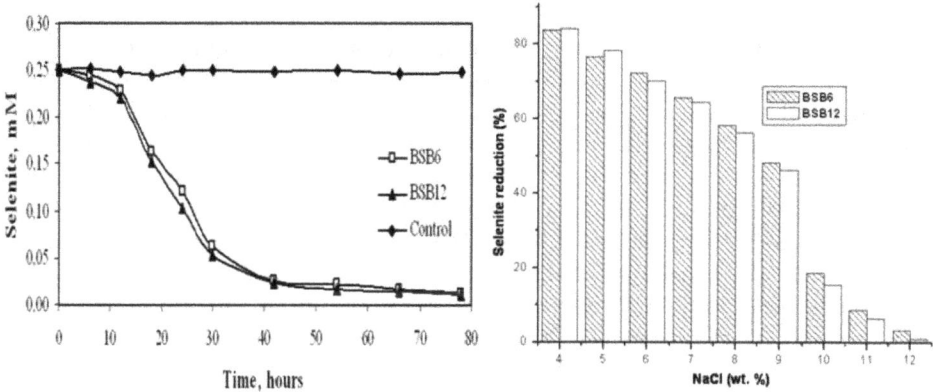

Fig. 63 (a) Selenite reduction by BSB6 and BSB12 at neutral pH, (b) Effect of NaCl on selenite reduction

4.5.3. Characterization of reduced elemental selenium

The bacterial cells associated with selenium particles, after 48 h incubation in presence of selenite (0.25 mM) at optimum pH (7.5)and temperature (37°C), were filtered through polycarbonate micropore filters (0.22 µm) and washed with Tris–HCl buffer (pH - 8.0) three times and fixed with 3% glutaraldehyde in 0.1 M phosphate buffer (pH -7.4) for 60 min. The suspension was centrifuged at 10,000g for 10 min and the pellet was washed with Tris–HCl buffer followed by deionized water three times. The samples were dehydrated with 70% ethanol, mounted on an aluminium stub, coated with gold and examined under JEOL (840A, Japan) SEM at 200 kV. Analysis of elemental selenium was also carried out simultaneously by energy dispersive microanalysis (EDX). For TEM, the 48 h culture grown in the presence of 0.25 mM selenite was centrifuged, washed three times with phosphate buffer saline (pH 7.4) followed by re-suspension in 1 mL of fixative solution (2.5% glutaraldehyde and 2% formaldehyde) and kept for 6 h at 4 °C. The post fixation and fixative solutions were removed by centrifugation. The cell pellet was then washed five times with phosphate buffer saline (pH 7.4) and finally re-suspended in 1 mL of sodium phosphate buffer. Aliquots (5 mL) of the cell suspension were mounted on 400 mesh Cu grids and allowed to dry overnight in a desiccator. Microscopic imaging of whole-mount cell was performed in a FEI Philips Morgagni 268 D Transmission Electron Microscope at 200 kV accelerating voltage. Confocal Laser Scanning Microscopy (CLSM) was carried out using a confocal microscope (Carl Zeiss, LSM7MP, Germany). For this, one drop of bacterial strain was put on the glass plate to get a biofilm. 1.0 µL of fluorescent solution (Rhodamin B) was carefully applied on top of the biofilm. After 30 min of incubation at room temperature in dark, the excess staining solution was removed by rinsing four times with phosphate buffer solution. Images were recorded at an excitation wavelength of 488 nm and an emission wavelength of 515 nm. Powder XRD pattern of reduced product was recorded on a Phillips PW3710 X-ray diffractometer using CuKa radiation at a scanning speed of 2 (2h) minute while the UV–Visible diffuse reflectance spectra (UV–Vis-DRS) were recorded on a Shimadzu UV–Visible spectrophoto meter (UV 2240) using $BaSO_4$ white reference.

The form of Se (0) is essential for the development of a suitable remediation process. As such attempts have been made to characterize the reduced product, associated with the cell, by SEM-EDX, TEM, XRD, UV–Vis-DRS and confocal images. SEM microphotographs of the elemental selenium with associated bacterial cells obtained after 48 h incubation in 0.25 mM. Se(IV) revealed the formation of Se(0) particles in and around the elongated bacterial cell of 1000 nm size (**Fig. 64a**). The EDX analysis of these particles displayed the characteristic SeLa absorption peak at 1.37 keV for Se (0) (**Fig. 64b**). TEM images (**Fig. 64c**) also showed the formation of selenium particles in the nanometric range associated with bacterial cells. Further analysis by TEM (**Fig. 64d**) also indicated the presence of extracellular spherical selenium nanoparticles of size 200 nm.

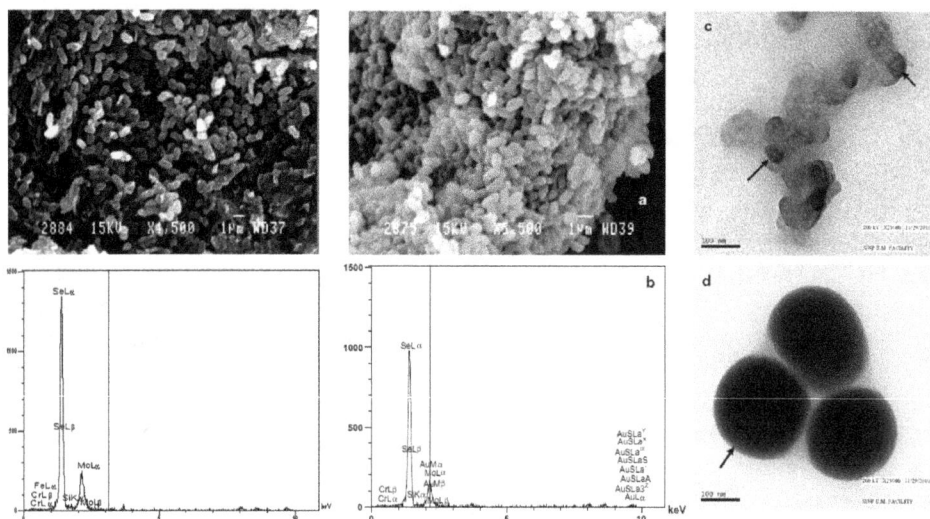

Fig 64: **SEM and EDX analysis of bacterial reduced selenite products**

4.6.4. Reduction of chromium

High concentrations of toxic levels of hexavalent chromium in the environment are of great concern worldwide (Losi *et al.*, 1994). Various industrial activities such as tannery, agro-food production, chemical manufacturing and oil and gas production generate saline wastewaters with high concentration of Cr(VI). Hexavalent chromium is toxic to all forms of life including humans and exhibits mutagenic, teratogenic and carcinogenic effects on biological systems due to its strong oxidizing nature (Cheung and Gu, 2007). Therefore, the concentration of Cr(VI) in the effluents needs to be reduced to the permissible limit (e.g. <0.05 mg L–1 as per US-EPA) using appropriate technology before being discharged into the environment (EPA, 1998). Bioremediation of Cr(VI) using Cr-resistant bacteria (CRB) provides a safe, effective and alternative viable process (Bai *et al.*, 2008). The potential of several Cr(VI) resistant bacterial strains, isolated from Cr(VI) free or contaminated soils, for reduction of Cr(VI) to less toxic Cr(III) under normal conditions have already been demonstrated previously (Nepple *et al.*, 2000; Dhal *et al.*, 2010). However, bioremediation of hexavalent chromium in industrial effluents and wastewater containing high salt concentration, especially in presence of NaCl, is more difficult as the biological treatment is strongly inhibited by the salts (Lefebvre *et al.*, 2006). In this context, the halophilic and halotolerant microorganisms may be suitable for bioremediation of such environment, since high concentration of anions and cations are needed for their growth (Margesin *et al.*, 2001). The present study was carried out to isolate and characterize high Cr(VI) resistant halophilic bacterial strain, from mangrove soils of Bhitarkanika, Odisha. using biochemical methods and 16S rRNA gene sequencing analysis. The kinetic of Cr(VI) reduction and characterization of reduced product were also investigated to understand the mechanisms of Cr(VI) resistance and reduction by halophilic bacteria. The findings of the present study will be a new report in deciphering the environmental significance of *Vigribacillus* sp., ahalophilic bacteria in bioremediation of toxic Cr(VI) from saline environment.

The reduction of Cr(VI) with the selected strain (H4) was carried out under varying incubation period, initial Cr(VI) concentration (50–250 mg L–1), pH (4–9), temperature (25–40 °C), salt concentration (6–12 wt.% NaCl), shaking speed (90–120 rpm) and commonly used carbon source with different structures and reducing abilities (glucose, fructose, sucrose, lactose, sodium acetate (1.0 wt.%) to optimize the parameters. All the reduction experiments were carried out in LB broth using pre-grown cells of H4 strain except otherwise mentioned. At pre-determined time intervals, 2.0 mL aliquots were drawn from the flasks, centrifuged (6000 × g for 10 min at 10 °C) and Cr(VI) was measured in the supernatant. All the experiments including controls were carried out at least in duplicate using an orbital motion shaker, and the average values with errors are reported (Fig. 65).

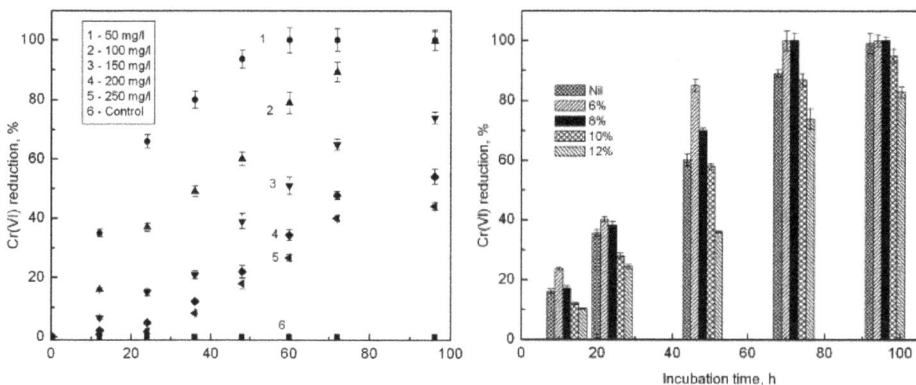

Fig 65: Effect of initial chromium concentration and NaCl on reduction of Cr(VI) at 35°C and pH 8.0.

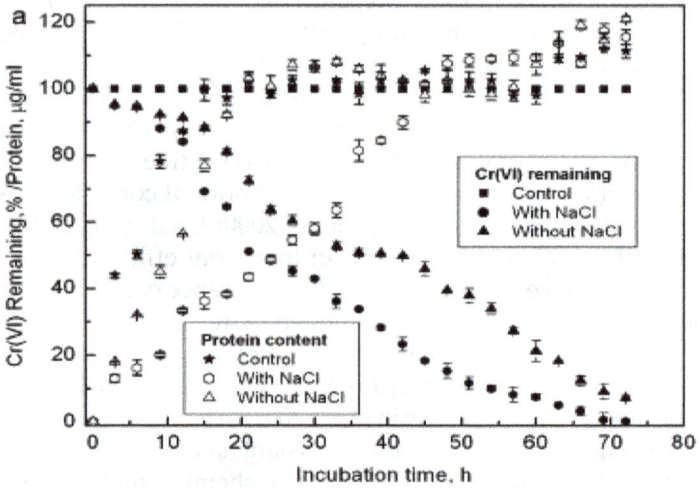

Fig 66: Simultaneous determination of Cr(VI) reduction and estimation of cellular protein content at 35°C, pH 8.0 with 100 mg L^{-1} Cr(VI) concentration

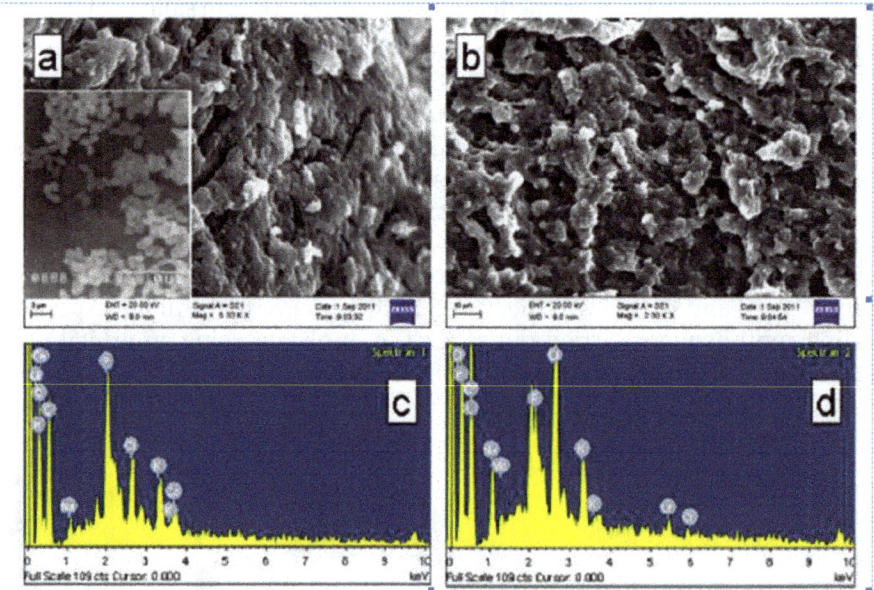

Fig 67: SEM (a and b) and EDS (c and d) images of Cr(III) species associated with *Vigribacillus* sp. H4.

4.6.4.1. Characterization of reduced products

In order to shed further light on the nature of reduced product, the product associated with cells was characterized by powder Xray diffraction, FT-IR, UV–visible diffuse reflectance (UV–vis–DRS), electron microscopy (SEM-EDS and TEM). For XRD, FT-IR and UV–vis–DRS, the bacterial cells associated with reduced product from a typical experiment with 100 mg L^{-1} Cr(VI), incubated for 72 h under optimized conditions, was separated by centrifugation at 10,000 ×g and 10 °C for 10 min. The supernatant liquid was discarded while the pellets were washed with deionized water and dried in air at 50 °C overnight. The bacterial cells grown under similar conditions but without Cr(VI), were also separated, dried and used for comparison. The XRD was recorded on a Rigaku Miniflex II X-ray diffractometer using Ni-filtered Cu K_ source. The FT-IR spectra of dried cells associated with or without reduced product in KBr phase were recorded using a Shimadzu, IR Affinity 1 FT-IR spectrophotometer. The UV–vis DRS were recorded on a Shimadzu, UV 2240 UV–visible spectrophotometer using BaSO$_4$ as reference. For SEM, the bacterial cell associated with reduced product, after 72 h incubation in the presence of 100 mg L^{-1} Cr(VI) was filtered through micro pore filtration unit and washed with phosphate buffer (pH 8.0) several times and fixed in 3% aqueous glutaraldehyde. It was then washed several times with Tris–HCl buffer followed by deionized water. The sample was then dried with ethanol under ambient conditions, mounted on a aluminium stub and coated with gold before it was examined by SEM (Jeol-840A, Japan) operated at 200 kV accelerating voltage. Analyses of desired elements were also carried out simultaneously by EDS. SEM of bacterial cells, grown without Cr(VI), were taken in an identical manner for comparison. For TEM, similarly cultured bacteria as that of SEM was centrifuged, washed three times with phosphate buffer saline (pH 7.4) followed by resuspension in 1 mL of fixative solution (2.5% glutaraldehyde and 2% formaldehyde) and kept for 6 h at 4 °C. The fixative solutions were removed by centrifugation. The cell pellet was then washed five times with phosphate buffer saline (pH 7.4) and finally re-suspended in 1 mL of sodium phosphate buffer. Aliquots (5 L) of the cell suspension were mounted on 400 mesh Cu grids and allowed to dry overnight in a desiccator. Microscopic images of whole-mount cells were taken on a FEI Philips Morgagni 268 D Transmission Electron Microscope operated at 200 kV accelerating voltage.

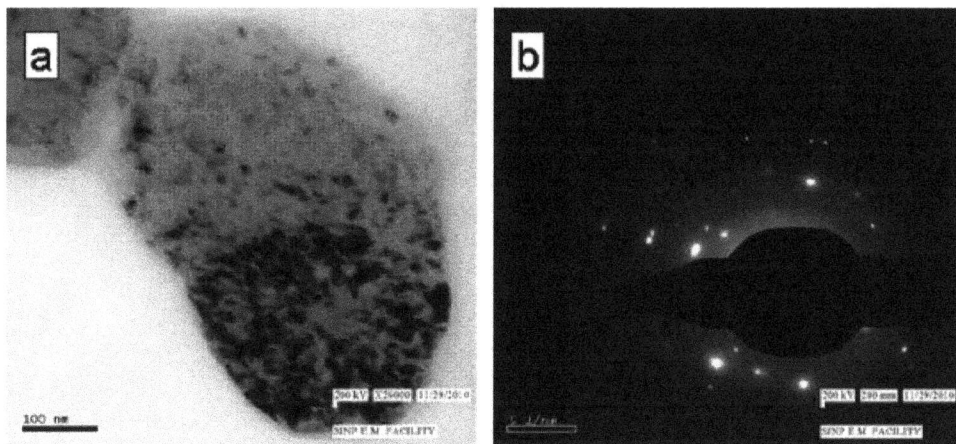

Fig 68: TEM (a)/SAED (b) images of Cr(III) species associated with Vigribacillus sp. H4.

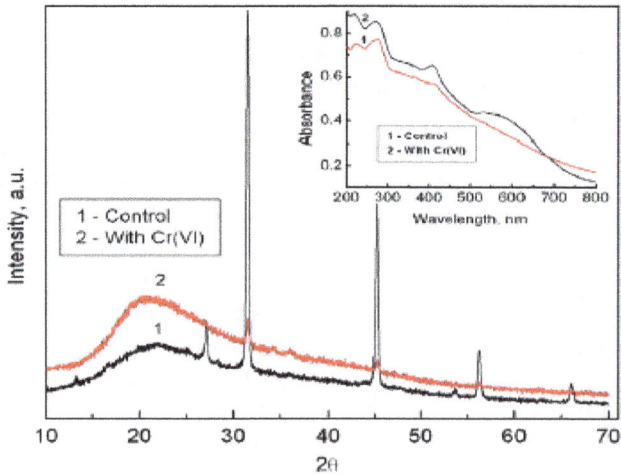

Fig 69: X-ray diffraction patterns and UV–visible diffuse reflectance spectra (inset) of reduced Cr(III) species associated with Vigribacillus sp. H4.

Conclusion

Mangrovee is the rich resorvor of microrganisms. All kinds of microbes such as algae, fungi and bacteria are found in these ecosystems which exhibit salt tolerance enzymes, antibiotics and various tress tolerant genes. Neverthless, they are less explored and their biotechnological applications are not much studied. Thus, the evaluation of biotechnological potential of mangrove microrganisms exhibiting salt tolerance capacity are extensively studied and documented. The novel biotechnological applications of many microrganisms exhibit great potential for their pharmaceutical, agricultural and industrial applications.

REFERENCES

Abdel-Wahab, M.A. 2005. Diversity of higher marine fungi from Egyptian Red Sea mangroves. *Bot. Mar.* 48: 348–355.

Acharya, S. and Chaudhury, A. 2011. Effect of nutritional and environmental factors on cellulase activity by thermophillic bacteria isolated from hot spring. *J. Sci. Ind. Res.* 70: 142–148.

Aksornkoae, S., Maxwell G.S., Havanond, S., Panichsuko 1992 plants in mangroves pub chalongrat co Ltd., 99 Tiemruammitr Rd, Huaykhwang, Bangkok, 10310, Thiland, ISBN: 97089011 7-3 (120pp).

Alias, S.A., Kuthubutheen, A.J. and Jones, E.R.G. 1995. Frequency of occurrence of fungi on wood in Malaysian mangroves.*Hydrobiol.* 295: 97–106.

Alongi, D.M. (1988). Bacterial productivity and microbial biomass in tropical mangrove sediments. Microb. Ecol. 15: 59-79.

Alongi, D.M. (2005). Mangrove–microbe–soil relations. In: Kristensen E, Haese RR, Kostka JE (eds) Interactions between macro- and microorganisms in marine sediments. American Geophysical Union, Washington, DC, pp. 85-103.

Alongi, D.M. 1989. The role of soft-bottom benthic communities in tropical mangrove and coral reef ecosystems. *Rev. Aquat. Sci.* 1:243–280.

Alongi, D.M. 1994. Zonation and seasonality of benthic primary production and community respiration in tropical mangrove forests. *Oecologia*. 98: 320-327.

Alongi, D.M. 1996. The dynamics of benthic nutrient pools and fluxes in tropical mangrove forests. *J.Mar. Res.* 54: 123-148.

Alongi, D.M. and Mukhopadhyay, S.K. 2014. Contribution of mangroves to coastal carbon cycling in low latitude seas. *Agric. For. Meteorol.* doi:10.1016/j.agrformet.2014.10.005

Alongi, D.M., 2008. Mangrove forests: Resilience, protection from tsunamis, and responses to global climate change. Estuar. Coast. Shelf Sci. 76, 1–13.

Alongi, D.M., Boto, K.G., Robertson, A.I. 1992. Nitrogen and Phosphorus cycles in tropical mangrove ecosystems. Washington DC: Am Geophys Univ 41:251–292.

Alongi, D.M., Christoffersen, P. and Tirendi, F. 1993. The influence of forest type on microbial-nutrient relationships in tropical mangrove sediments. *J. Exp. Mar. Biol. Ecol.* 171: 201-223.

Alongi, D.M., Ramanathan, A.L., Kannan, L., Tirendi, F., Trott, L.A. and Bala Krishna Prasad, M. 2005. Influence of human-induced disturbance on benthic microbial metabolism in the Pichavaram mangroves, Vellar-Coleroon estuarine complex, India. *Mar. Biol.* 147:1033-1044.

Alongi, D.M., Sasekumar, A., Tirendi, F. and Dixon, P. 1998. The influence of stand age on benthic decomposition and recycling of organic matter in managed mangrove forests of Malaysia. *J. Exp. Mar. Biol. Ecol.* 225:197–218.

Amann, R.I., Ludwig, W. and Schleifer, K.H. (1995) Phylogenetic identification and in situ detection of individual microbial cells without cultivation. *Microbiol. Rev.* 59:143-169.

Anderson, G. (1980). Assessing organic phosphorus in soils. In: Khasawneh, F.E., Sample, E.C., Kamprath, E.J. (Eds.). The Role of Phosphorus in Agriculture. Madison, Wis: Amer Soc Agronom pp. 411-32.

Andrews, T.J., Clough, B.F. and Muller, G.J. 1984. Photosynthetic gas exchange properties and carbon isotope ratios of some mangroves in North Queensland. In Physiology and management of mangroves (Teas, H.J. ed.) pp. 15-23 W Junk, The Hague.

Ara, I., Kudo, T., Matsumoto, A., Takahashi, Y. and Omura, S. 2007. *Nonomuraea maheshkhaliensis* sp. nov., a novel actinomycete isolated from mangrove rhizosphere mud. *J. Gen. Appl. Microbiol.* 53:159–166.

Armando, C.F.D., Andreote, F.D., Dini-Andreote, F., Lacava, P.T., Sá A.L.B.,Melo, I.S., Azevedo, J.L. and Araujo, W.L. 2009. Diversity and biotechnological potential of culturable bacteria from Brazilian mangrove sediment. *World. J. Microbiol. Biotechnol.* 25: 1305–1311.

Ashabil, A., Lutfiye, K. and Burhan, A. 2011. Alkaline thermostable and halophilic endoglucanase from *Bacillus licheniformis* C108. *Afr. J. Biotechnol.* 10:789–796.

Ayukai, T. & E. Wolanski. 1997. Importance of biologi-cally medicated removal of the sediments from the Fly River plume. Papua New Guinea. Estuar. Coast. Shelf Sci. 44: 629-639

Bai, H.J., Zhang, Z.M., Yang, G.E. and Li, B.Z. 2008. Bioremediation of cadmium by growing *Rhodobacter sphaeroides*: kinetic characteristic and mechanism studies, *Biores. Technol.* 99: 7716–7722.

Bakare, M.K., Adewale, I.O., Ajayi, A. and Shonukan, O.O. 2005. Purification and characterization of cellulase from the wildtype and two improved mutants of *Pseudomonas fluorescens. Afr. J. Biotechnol.* 4: 898–904.

Ball, M.C. and Pidsley, S.M. 1995. Growth responses to salinity in relation to distribution of two mangrove species, *Sonneratia alba* and *S. lanceolata*, in northern Australia. *Func. Ecol.* 9 (1) : 77-85.

Bano, N., Nisa, M-U., Khan, N., Saleem, M., Harrison, P.J., Ahmed, S.I. and Azam, F. 1997. Significance of bacteria in the flux of organic matter in the tidal creeks of the mangrove ecosystem of the Indus river delta, Pakistan. *Mar. Ecol. Prog. Ser.* 157: 1-12.

Baruah AD (2005) Point Calimere Wildlife and Bird Sanctuary-A Ramsar site. Tamil Nadu Forest Department, p. 180.

Bashan Y, Holguin G (1997) Azospirillum - plant relationships: environmental and physiological advances (1990 -1996). Can J. Microbiol 43: 103-121

Behera, B., Das, M. and Rana, G.S. 2012. Studies on ground water pollution due to iron content and water quality in and around, Jagdalpur, Bastar district, Chattisgarh, India. *J. Chem. Pharm. Res.* 4 (8): 3803–3807.

Behera, B.C., Mishra, R.R., Patra, J.K., Sadangi, K., Dutta, S.K. and Thatoi, H.N. 2013. Impact of heavy metals on bacterial communities from mangrove soils of the Mahanadi Delta (India). *Chem. Ecol.* http://dx.doi.org/10.1080/02757540 .2013.810719

Beijerinck, M.W. 1904. Phenomenes de reduction produits par les microbes (Conference avec demonstrations faite - Delft, le, *Arch. Neer. Sci. Ser.* 29: 131-157.

Benka-coker, M.O. and Olumagin, A. 1995. Waste drilling fluid utilising microorganisms in a tropical mangrove swamp oilfield location. *Biores. Technol.* 53(3):211–215

Bhatt JR, Kathiresan K (2011) Biodiversity of mangrove ecosystems in India. In: Towards conservation and management of mangrove ecosystem in India.

Bloom, H. and Ayling, G.M. 1977. Heavy metals in the Derwent Estuary. *Environ Geol.* 2: 3-22.

Borse, B.D. 1988. Frequency occurrence of mangrove fungi from Maharastra coast, India. *Indian J. Mar. Sci.* 17: 165–167.

Boto, K.G. 1982. Nutrient and organic fluxes in mangroves. In: Clough, B.F. (Ed.). Mangrove Ecosystem in Australia, Australian National University Press, Canberra, pp. 239-257.

Boto, K.G. and Robertson, A.I. 1990. The relationship between nitrogen fixation and tidal exports of nitrogenin a tropical mangrove system. *Est. Coast. Shelf. Sci.* 31: 531-540.

Boto, K.G., Bunt, J.S. and Wellington, J.T. 1984. Variation in mangrove forest productivity in northern Australia and Papua New Guinea. Estuarine, *Coast. Shelf Sci.* 19: 321-329.

Boto. K.G. and Wellington, J.T. 1983. Phosphorus and nitrogen nutritional status of a northern Australian mangrove forest. *Mar. Ecol. Prog. Ser.* 11: 63-69.

Bradford, M.M. 1976. A Rapid and Sensitive Method for the Quantitation of Microgram Quantities of Protein Utilizing the Principle of Protein-Dye Binding. *Analyt Biochem.* 72: 248-254.

Brito, E.M., Guyoneaud, R., Goñi-Urriza, M., Ranchou- Peyruse, A., Verbaere, A., Crapez, M.A.C. 2006. Characterization of hydrocarbonoclastic bacterial communities from mangrove sediments in Guanabara Bay, Brazil. *Res. Microbiol.* 157:752–762.

Camilleri, J.C. & Ribi, G. 1986. Leaching of dissolved organic carbon (DOC) from dead leaves, formation of flakes of DOC, and feeding on flakes by crustaceans in mangroves. *Mar. Biol.* 91: 337–344.

Canfield, D.E., Kristensen, E. and Thamdrup, B. 2005. Aquatic Geomicrobiology. Elsevier,Amsterdam.

Cha, J.M., Cha, W.S. and Lee, J.H. 1999. Removal of organosulphur odour compound by *Thiobacillus novellas* SRM, Sulphur oxidising microorganism. *Proc. Biochem.* 34: 659-665.

Chadha S and Kar C S (1999) Bhitarkanika: Myth and Reality. Natraj Publishers, Dehradun

Cheng, Z.S., Pan, J.H., Tang, W.C., Chen, Q.J. and Lin, Y.C. 2009. Biodiversity and biotechnological potential of mangrove-associated fungi. *J. For. Res.* 20(1): 63–72.

Cheung, K.H. and Gu, J.D. 2007. Mechanism of hexavalent chromium detoxification by microorganisms and bioremediation application potential: a review. *Int. Biodeter. Biodegrad.* 59: 8–15.

Choudhury, B.P. 1990. The unique mangrove forest of Bhitarkanika in the state of Orissa. *Orissa rev.* 46(9): 34-39.

Clough, B.F., Andrews, T.J. and Cowan, I.R. 1982. Physiological processes in mangroves. In : Mangrove ecosystems in Australia : structure, function and management (Clough, B.F. ed.). Australian National University Press, Canberra, pp. 193-210.

Coleman, J.E., 1992. Structure and mechanism of alkaline phosphatase. Annual review of Biophysics and Biomolecular Structure. 21,441–483.

Cordeiro-Marino, M., Braga, M.R.A., Eston, V.R., Fujii, M.T. and Yokoya, N.A. 1992. Mangrove macro algal communities in Latin America. The state of the art and perspectives. In: Seeliger U (ed) Coastal plant communities of Latin America. Academic, San Diego, pp 51–64.

Corredor, J.E., Morell, J.M., 1994. Nitrate depuration of secondary sewage effluents in mangrove sediments. *Estuaries.* 17: 295–300.

Costa, R.B., Silva, M.V.A., Freitas, F.C., Leitao, V.S.F., Lacerda, P.S.B., Ferrara, M.A. and Bon, E.P.S. 2008. Mercado e Perspectivas de Uso de Enzimas Industriais e Especiais no Brasil. In: Bon EPS, Ferrara MA, Corvo ML, Vermelho AB, Paiva CLA, Alencastro RB, Coelho RRR, editors. Enzimas em Biotecnologia, Producao, Aplicac oes e Mercados. 1st edn. Rio de Janeiro. *Interciencia.* p. 463–488.

Crib, A.B. and Crib, J.W. 1955. Marine fungi from Queensland 1. Univ Queensland Pap Dept Bot. 3:77–81.

Cundell, A.M., Brown, M.S., Stanford, R. and Mitchell, R. 1979. Microbial degradation of *Rhizophora mangle* leaves immersed in the sea. *Estuar. Coast. Mar. Sci.* 9: 281-286.

Cusack, D.F., Chou, W.W., Yang, W.H., Harmon, M.E., Silver, W.L. and Lidet, T. 2009. Controls on long- term root and leaf litter decomposition in neotropical forests. Glob *Change. Biol.* 15(5): 1339-1355.

D'Souza, D.T., Tiwari, R., Sah, A.K. and Raghukumar, C. 2006. Enhanced production of laccase by a marine fungus during treatment of colored effluents and synthetic dyes. Enzym Microb. Technol. 38: 504–511.

Dahdouh-Guebas, F., Kairo, J.G., Jayatissa, L.P., Cannicci, S., Koedam, N., 2002a. An ordination study to view vegetation structure dynamics in disturbed and undisturbed mangrove forests in Kenya and Sri Lanka. Plant Ecol. 161,123–135.

Dahdouh-Guebas, F., Zetterström, T., Ronnback, P., Troell, M., Wickramasinghe, A., Koedam, N., 2002b. Recent changes in land-use in the Pambala-Chilaw Lagoon complex (Sri Lanka) investigated using remote sensing and GIS: conservation of mangroves vs. development of shrimp farming. Environ. Dev. Sustainability 4, 185–200.

Dalal, R.C. 1977. Soil organic phosphorus. *Adv. Agron.* 29: 83-117.

Dar, S.A., Kleerebezem, R., Stams, A.J.M., Kuenen, J.G. and Muyzer, G. 2008. Competition and coexistence of sulfate-reducing bacteria, acetogens and methanogens in a lab-scale anaerobic bioreactor as affected by changing substrate to sulfate ratio. *Appl. Environ. Microbiol.* 78: 1045–1055.

Davis, S.E., Coronado-Molina, C., Childers D.L. and Day, J. 2003. Temporally dependent C, N, and P dynamics associated with the decay of *Rhizophora mangle* L. leaf litter in oligotrophic mangrove wetlands of the southern Everglades. *Aq. Bot.* 75: 199-215.

Desai, C., Pathak, H. and Madamwar D. 2010. Advances in molecular and "omics" technologies to gauge microbial communities and bioremediation at xenobiotic/ anthropogen contaminated sites. *Biores. Technol.* 101(6): 1558–1569.

Dhal, B., Thatoi, H.N., Das, N.N. and Pandey, B.D. 2010. Reduction of hexavalent chromium by *Bacillus* sp. isolated from chromite mine soils and characterization of reduced product. *J. Chem. Technol. Biotechnol.* 85: 1471–1479.

Dias, A.C.F., Andreote, F.D., Dini-Andreote, F., Lacava, P.T., Sa, A.L.B., Melo, I.S., Azevedo, J.L. and Araujo, W.L. 2009. Diversity and biotechnological potential of culturable bacteria from Brazilian mangrove sediment. *World J. Microbiol. Biotechnol.* 25(7): 1305–1311.

Dien, B.S., Ximenes, E.A., O'Bryan, P.J., Moniruzzaman, M., Li, X.-L., Balan, V., Dale, B., Cotta, M.A., 2008. Enzyme characterization for hydrolysis of AFEX and liquid hot-water pretreated distillers' grains and their conversion to ethanol. Bio-resour. Technol. 99, 5216-5225.

D'Souza, D.T., Tiwari, R., Sah, A.K. and Raghukumar, C. 2006. Enhanced production of laccase by a marine fungus during treatment of colored effluents and synthetic dyes. Enzym Microb. Technol. 38: 504–511.

Dittmar, T., Hertkorn, N., Kattner, G. and Lara R. J. 2006. Mangroves, a major source of dissolved organic carbon to the oceans. *Glob Biogeochem. Cyc.* 20: GB1012. doi:10.1029/2005GB002570.

Duke, N., Meynecke, J., Dittmann, S., Ellison, A., Anger. K, U. Berger, Cannicci, S., Diele, K., Ewel, K., Field, C. Koedam, N., Lee, S., Marchand, C., Nordhaus, I., Dahdouh-Guebas, F., 2007. A world without mangroves. Science 317(5834),41-42.

Eccleston, G.P., Brooks, P.R. and Kurtböke, D.I. 2008. The occurrence of bioactive micromonosporae in aquatic habitats of the Sunshine Coast in Australia. *Mar. Drug.* 6: 243-261.

Environmental Protection Agency (EPA), Toxicological Review of Hexavalent Chromium, CASNR, 18540-29-9, Washington DC, USA, 1998.

Ewel KC, Twilley RR, Ong JE (1998) Different kinds of mangrove forests provide different goods and services. Global Ecol Biogeog Let 7: 83–94.

Farnsworth, E. J., and A. M. Ellison. 1997. Global patterns of pre-dispersal propagule predation in mangrove forests. *Biotropica* 29: 318-330.

Field CB, Osborn JG, Hoffman LL, Polsenberg JF, Ackerly DD, et al. (1998) Mangrove biodiversity and ecosystem function. Global Ecol Biogeog Let 7: 3–14.

Fogel, M.L., Wooller, M.J., Cheeseman, J., Smallwood, B.J., Roberts, Q., Romero, I. and Jacobson Meyers, M. 2008. Unusually negative nitrogen isotopic compositions (15N) of mangroves and lichens in an oligotrophic, microbially-influenced ecosystem. *Biogeosci. Discuss.* 5: 937 –969.

Fordyce, F.M. 2005. Selenium deficiency and toxicity in the environment. In: Selinus, O., Alloway, B., Centeno, J.A., Finkelman, R.B., Fuge, R., Lindh, U., Smedley, P. (Eds.), Essentials of Medical Geology. Elsevier Academic Press, Amsterdam, Holland, pp. 373–416.

FSI 2015. India State of Forest Report 2015. Forest Survey of India, Ministry of Environment and Forests, Government of India, pp.35-39.

Furukawa, K., Wolanski, E. & Mueller, H. 1997. Currents and sediments transport in mangrove forests. *Estuar Coast Shelf Sci*, 44: 301-310

Gao, J.H., Weng Zhu, D., Yuan, M., Guan, F. and Xi, Y. 2008. Production and characterization of cellulolytic enzymes from the thermo-acidophilic fungas *Aspergillus terreus* M11 under solid-state cultivation of corn stover. *Biores. Technol.* 99:7623–7629.

Garrettson–Cornell, L. and Simpson, J. 1984. Three new marine *Phytophthora* species from New South Wales. *Mycotax*.19: 453–470.

Gayathri, S., Saravanan, D., Radhakrishnan, M., Balagurunathan, R. and Kathiresan, K. 2010. Bioprospecting potential of fast growing endophytic bacteria from leaves of mangrove and salt-marsh plant species. *Ind. J. Biotech.* 9: 397-402.

George, P.S., Ahmad, A. and Rao, M.B. 2001. Studies on carboxymethyl cellulase produced by an alkalothermophilic actinomycete. *Biores. Technol.* 77:171–175.

Giani, L., Bashan, Y., Holguin, G. and Strangmann, A. 1996. Characteristics and methanogenesis of the Balandra lagoon mangrove soils, Baja California Sur, Mexico. *Geoderm.* 72: 149-160.

Gilman, E., Ellison, J., Duke, N. and Field, C., 2008. Threats to mangroves from climate change and adaptation options: a review. Aquat. Bot. 89(2), 237–250.

Giri, C., Ochieng, E., Tieszen, L.L., Zhu, Z., Singh, A., Loveland, T., Masek, J. (2011). Status and distribution of mangrove forests of the world using earth observation satellite data. Glob, Eco. Biogeograph. 20: 154-159.

Goldstein, A.H. 1994. Involvement of the quinoprotein glucose dehydrogenase in the solubilization of exogenous phosphates by gram-negative bacteria. In: Torriani-Gorini A, Yagil, E., Silver, S., editors. Phosphate in microorganisms: Cellular and Molecular Biology. Washington, DC: ASM Press. 197-203.

Gopal, B. and Krishnamurthy, K. 1993. Wetlands of South Asia. In Wetlands of the world 1:Inventory, ecology and management, eds DF Whigham, D Dykyjova & S Hejny, Kluwe Academic Publishers, 345–414.

Gulve, R.M. and Deshmukh, A.M. 2011. Enzymatic activity of actinomycetes isolated from marine sediments. *Recent Res. Sci. Technol.* 3(5): 80–83.

Gupta, N., Das, S. and Basak, U.C. 2007. Use of extracellular activity of bacteria isolated from Bhitarkanika mangrove ecosystem of Orissa coast. *Malay. J. Microbiol.* 3(2): 15-18.

Gupta, N., Mishra, S. and Basak, U.C. 2009. Microbial population in phylosphere of mangroves grow in different salinity zones of Bhitarkanika (India). *Acta Botanic Malact.* 34: 1-5.

Gupta, R., Malik, A., Rizvi, M. and Ahmed, M. 2016. An Alarming Increase of Fungal Infections in Intensive Care Unit: Challenges in the Diagnosis and Treatment. *J. Appl. Pharmaceutic. Sci.* 6 (11): 114-119.

Gyaneshwar, P., Naresh Kumar, G. and Parekh, L.J. 1998. Effect of buffering on the P-solubilizing ability of microorganisms. *World J. Microbiol. Biotechnol.* 14: 669-673.

Haferburg, G. and Kothe, E. 2007. Microbes and metals: interactions in the environment. *J Basic. Microbiol.* 47(6): 453-67.

Haight, M. 2005. Assessing the environmental burdens of anaerobic digestion in comparison to alternative options for managing the biodegradable fraction of municipal solid wastes. *Water Sci. Technol.* 52: 553–559.

Haines, H.H. (1921-25): Hie Botany of Bihar and Orissa, 6 parts. London. Bot. Survey of India, Calcutta (Rep. Edn. 1961).

Harbison, P. 1986. Mangrove mud a sink and a source for trace metals. *Mar. Poll. Bullet.* 17: 246-250.

Harley, J.L. and Smith, S.E. 1983. Mycorrhizal symbiosis. London, New York Academic Press.

Hesse, P.R. 1962. Phosphorus fixation in mangrove swamp muds. *Nature.* 193: 295-296.

Hicks, B.J. and Silvester, W.B. 1985. Nitrogen fixation associated with the New Zealand mangrove *Avicennia marina* (Forsk) *Vierh. Var. resinifera* (Forst. F) *Bakh. Appl. Environ. Microbiol.* 49: 955-959.

Hirano, T., Kurosawa, H., Nakamura, K. and Amano, Y. 1996. Simultaneous removal of hydrogen sulphide and trimethylamine by a bacterial deodorant. *J. Ferment. Bioengineer.* 81: 337-342.

Hoffmann, L. (1999). Marine cyanobacteria in tropical regions: Diversity & Ecology. *Eur. J. Phycol.* 34: 371-379.

Holguin, G., Guzman, M.A. and Bashan, Y. 1992. Two new nitrogen fixing bacteria from the rhizosphere of mangrove trees: their isolation, identification and *in vitro* interaction with rhizosphere *staphylococcus* sp. *FEMS Microbiol.* 101: 207-216.

Holguin, G., Vazquez, P. and Bashan, Y. 2001. The role of sediment microorganisms in the productivity, conservation and rehabitation of mangrove ecosystems: an overview. *Biol. Fertil. Soil.* 33: 265-278.

Holmboe, N. and Kristensen, E. 2002. Ammonium adsorption in sediments of a tropical mangrove forest (Thailand) and a temperate Wadden Sea area (Denmark). *Wetlands Ecol. Manag.*10: 453-460.

Holmer, M. and Storkholm, P. 2001. Sulphate reduction and sulphur cycling in lake sediments: a review. *Freshwater Biol.* 46: 431-451.

Holmer, M., Kristensen, E., Banta, G., Hansen, K., Jensen, M.H. and Bussawarit, N. 1994. Biogeochemical cycling of sulfur and iron in sediments of a southeast Asian mangrove, Phuket Island, Thailand. *Biogeochem.* 26: 145-161.

Hong, K., Gao, A.H., Xie, Q.Y., Gao, H., Zhuang, L., Lin, H.P., Yu, H.P., Li, J., Yao, X.C., Goodfellow, M. and Ruan, J.S. 2009. Actinomycetes for marine drug discovery isolated from mangrove soils and plants in China. *Mar. Drug.* 7: 24–44.

Huang, H., Feng, X., Xiao, Z., Liu, L., Li, H., Ma, L., Lu, Y., Ju, J. She, Z. and Lin, Y. 2011. Azaphilones and p-terphenyls from the mangrove endophytic fungus *Penicillium chermesinum* (ZH4-E2) isolated from the South China Sea. *J. Nat. Prod.* 74(5): 997–1002.

Huang, H., Lv, J., Hu, Y., Fang, Z., Zhang, K. and Bao, S. 2008. *Micromonospora rifamycinica* sp. nov, a novel actinomycete from mangrove sediment. *Int. J. Syst. Evol. Microbiol.* 58: 17–20.

Hyde KD. 1990. A comparison of the intertidal mycota of five mangrove tree species. *Asi Mar. Biol.* 7:93–107.

Hyde, K.D. 1996. Marine fungi. In: Grurinovic C, Mallett K, editors. Fungi of Australia. Vol 1B. Canberra: ABRS/CSIRO; p.39–64.

Hyde, K.D., Jones, E.B.G., Leano, E., Pointing, S.B., Poonyth, A.D. and Vrijmoed, L.L.P. 1998. Role of fungi in marine ecosystems. *Biodiver. Conserv.* 7: 1147–1161.

Isaka, M., Suyarnsestakorn, C., Tanticharoen, M. 2002. Aigialomycins A–E, new resorcylic macrolides from the marine mangrove fungus *Aigialus parvus*. *J. Org. Chem.* 67:1561–1566.

Janssen, A.J.H., Lettinga, G. and de Keizer, A. 1999. Removal of hydrogen sulphide from wastewater and waste gases by biological conversion to elemental sulphur: colloidal and interfacial aspects of biologically produced sulphur particles. *Colloids Surf.* 151: 389-397.

Jennerjahn, T. C. and Ittekkot V. 2002, Relevance of mangroves for the production and deposition of organic matter along tropical continental margins, *Naturwissenschaften*, 89: 23–30.

Jiménez, J.A. 1990. The structure and function of dry weather mangroves on the Pacific coast of Central America, with emphasis on *Avicennia bicolour* forests. *Estuaries*. 13 (2): 182-192.

Joseph, I. and Paul Raj, R. 2007. Isolation and characterization of phytase producing *Bacillus* strains from mangrove ecosystem. *J. Mar. Biol. Assoc. Ind.* 2: 177–182.

Kar M, Mishra D (1976). Catalase, peroxidase, and polyphenoloxidase activities during rice leaf senescence. Plant Physiol. 57:315-319.

Kashiwa, M., Nishimoto, S., Takahashi, K., Ike, M. and Fujita, M. 2000. Factors affecting soluble selenium removal by a selenate reducing bacterium *Bacillus* sp. SF-1. *J. Biosci. Bioeng.* 89: 528–533.

Kathiresan K, Saravanakumar K, Anburaj R, Gomathi V, Abirami G, Sahu SK, Anandhan S (2011) Microbial enzyme activity in decomposing leaves of mangroves. Int J Adv Biotechnol Res 2(3):382-389.

Kathiresan, K. 2003. How do mangrove forests induce sedimentation. *Rev. Biol. Trop.* 51: 355–360.

Kathiresan, K. and Bingham, B.L. 2001. Biology of mangroves and mangrove ecosystem. *Adv. Mar. Biol.* 40: 81-251.

Kathiresan, K. and Faisal, A.M. 2006. Managing Sundarbans for uncertainty and sustainability. International Conference and Exhibition on mangroves of Indian and Western Pacific Oceans, ICEMAN, Kuala Lumpur, Malaysia, pp. 1-31.

Kathiresan, K. and Qasim, S.Z. 2005. Biodiversity of mangrove ecosystems. Hindustan, New Delhi, p 51

Kathiresan, K., Rajendran, N. and Thangadurai, G. 1996. Growth of mangrove seedlings in intertidal area of Vellar estuary southeast coast of India. *Ind. J. Mar. Sci.* 25: 240-243.

Kaur, J., Chadha, B.S., Kumar, B.A. and Saini, H.S. 2007. Purification and characterization of two endoglucanases from *Melanocarpus* sp. MTCC 3922. *Biores. Technol.* 98: 74–81.

Ke, L., Wang, W.Q., Wong, T.W.Y., Wong, Y.S. and Tam N.F.Y. 2003. Removal of pyrene from contaminated sediments by mangrove microcosms. *Chemosphere*. 51: 25–34.

Khan, J.A. and Kumar, D. 2012. Production and partial purification of cellulase from bacteria inhabiting cow dung. *Res. J. Pharm. Biol. Chem. Sci.* 3: 547–558.

Kishore, P. 2011. Isolation, characterization and identification of Actinobacteria of Mangrove ecosystem, Bhitarkanika, Odisha. MSc thesis submitted to National Institute of Technology, Odisha, India .

Kohlmeyer, J. & E. Kohlmeyer. 1979. Marine mycology. Then higher fungi. Academic, New York, USA. 690 p.

Kolmert A., Wikström, P. and Hallberg, K.B. 2000. A Fast and Simple Turbidimetric Method for the Determination of Sulfate in Sulfate-Reducing Bacterial Cultures. *J. Microbiol. Method. 41: 179-184.*

Komiyama, A., Ong, J.E. and Poungparn, S. 2008. Allometry, biomass, and productivity of mangrove forests: a review. Aquatic Bot. 89: 128-137.

Kothamasi, D., Kothamasi, S., Bhattacharyya, A., Kuhad, R.C., Babu, C.R. 2006. Arbuscular mycorrhizae and phosphate solubilising bacteria of the rhizosphere of the mangrove ecosystem of Great Nicobar island. *Ind. Biol. Fertil. Soil.* 42: 358-361.

Krauss KW, Keeland BD, Allen JA, Ewel KC, Johnson DJ. 2007. Effects of season, rainfall, and hydrogeomorphic setting on mangrove tree growth in Micronesia. Biotropica 39: 161–170.

Kreuzwieser, J., Buchholz, J. and Rennenberg, H. 2003. Emission of methane and nitrous oxide by Australian mangrove ecosystems. *Plant Biol*, 5(4): 423-431. http://dx.doi.org/10.1055/s-2003-42712

Kristensen, E., Bouillon, S., Dittmar, T. and Marchand C. 2008. Organic matter dynamics in mangrove ecosystems, *Aquat. Bot.* doi:10.1016/j. aquabot.2007.12.005.

Kristensen, E., Holmer, M. and Bussarawit, N. 1991. Benthic metabolism and sulfate reduction in a south-east Asian mangrove swamp. *Mar. Ecol. Prog. Ser.* 73: 93-103.

Kristensen, P., Judge, M.E., Thim, L., Ribel, U., Christjansen, K.N., Wulff, B.S., Clausen, J.T., Jensen, P.B., Madsen, O.D., Vrang, N., Larsen, P.J. and Hastrup, S. 1998. Hypothalamic CART is a new anorectic peptide regulated by leptin. *Nature.* 393:72-76.

Kuhad, R.C., Gupta, R. and Singh, A. 2011. Microbial cellulases and their industrial applications. *Enz Res.* 2011: 280696. doi:10.4061/2011/280696.

Kushner, D.J. 1985. The halobacteriaceae. In: The bacteria, vol. 8. Academic Press, New York, pp 171–214.

Lacerda L.D., Carvalho C.E.V., Tanizaki K.F., Ovalel A.R.C. and Rezende C.E. 1993. The biogeochemistry and trace metals distribution of mangrove rhizospheres. *Biotropica,* 25: 252-257.

Laemmli, U. K. 1970. Cleavage of structural proteins duringthe assembly of the head of bacteriophage T4. *Nature.* 227: 680---685.

Lakshmanaperumalsamy, P. 1987. Nitrogen fixing bacteria, *Azotobacter sp.* in aquatic sediment. *Fish Technol. Soc.* 24(2): 126-128.

Laksmanaperumalsamy, P., Chandramohan, D. and Natarajan, R. 1978. Antibacterial and antifungal activity of streptomycetes from Porto Novo coastal environment. *Mar. Biol.* 11:15–24.

Lee, R.Y. and Joye, S.B. 2006. Seasonal patterns of nitrogen fixation and denitrification in oceanic mangrove habitats. *Mar. Ecol. Progress Series.* 307: 127-141.

Lefebvre, O. and Moletta, R. 2006. Treatment of organic pollution in industrial saline wastewater: a literature review. *Water Res.* 40: 3671–3682.

Liang, J.B., Chen, Y.Q., Lan, C.Y., Tam, F.Y., Zan, Q.J. and Huang, L.N. 2006. Recovery of novel bacterial diversity from mangrove sediment. *Mar. Biol.* 150: 739-747.

Lin, G.H. and Sternberg, L.D.S.L. 1993. Effects of salinity fluctuation on photosynthetic gas exchange and plant growth of the red mangrove (*Rhizophora mangle* L.). *J. Experiment. Bot.* 44 (258): 9-16.

Lin, Y.C. and Zhou, S.N. 2003. Marine microorganism and its metabolites. Chemical Industry, Beijing, pp 426–427.

Lin, Y.C., Wu, X.Y., Deng, Z.J., Wang, J., Zhou, S.N., Vrijmoed, L.L.P. and Jones, E.B.G. 2002b. The metabolites of the mangrove fungus *Verruculina enalia* No. 2606 from a salt lake in the Bahamas. *Phytochem.* 59: 469–471.

Lin, Y.C., Wu, X.Y., Feng, S., Jiang, G.C., Luo, J.H., Zhou, S.N., Vrijmoed, L.L.P., Jones, E.B.G., Krohn, K., Steingröver, K. and Zsila, F. 2001. Five unique compounds: xyloketals from mangrove fungus *Xylaria* sp. from the South China Sea coast. *J. Org. Chem.* 66: 6252–6256.

Lin, Y.C., Wang, J., Wu, X.Y., Zhou, S.N., Vrijmoed, L.L.P. and Jones, E.B.G. 2002a. A novel compound enniatin G from the mangrove fungus *Halosarpheia* sp. (strain 732) from the South China Sea. *Aust. J. Chem.* 55: 225–227.

Liu, A.R., Wu, X.P. and Tong, X.U. 2007. Research advances in endophytic fungi of mangrove. *Chin J. Appl. Ecol.* 18: 912–918.

Loka Bharathi, P.A., Oak, S. and Chandramohan, D. 1991. Sulfate-reducing bacteria from mangrove swamps II: their ecology and physiology. *Oceanol. Acta.* 14: 163-171.

Losi, M.E., Amrhein C. Frankenberger, W.T. 1994. Environmental biochemistry of chromium. *Rev. Environ. Contam. Toxicol.* 136: 91–121.

Lowry, O. H., Rosebrough, N. J., Farr, A. L. & Randall, R. J. 1951. Protein measurement with the Folin phenol reagent. *J. Biol. Chem.* 193: 265---275.

Lu, C. Y., Wong, Y. S., Tam, N. F. Y., Ye, Y. and Lin, P. 1999. Methane flux and production from sediments of a mangrove wetland on Hainan Island China. *Mang. Salt Marsh.* 3: 41–49

Lu, W.J., Wang, H.T, Yang, S.J., Wang, Z.C. and Nie, Y.F. 2005. Isolation and characterisation of mesophilic cellulose degrading bacteria from flower stalks-vegetable waste co-composting system. *J. Gen. Appl. Microbiol.* 51: 353-360.

Lugo, A. E.; Snedaker, S.C. 1975. Properties of a Mangrove Forest in Southern Florida. In: G. Walsh, S. Snedaker and H. Teas (eds.) International Symp. Biology and Management of Mangroves, Vol. I, pp. 170-212. Gainesville, University of Florida.

Lugo, A. E. and Snedaker. S. C. 1974. The ecology of mangroves. *Ann. Rev. Ecol. System.* 5: 39–64.

Lyimo, T.J., Pol, A. and Op den Campa, J.M. 2002. Methane emission, sulphide concentration and redox potential profiles in Mtoni mangrove sediment, Tanzania. West Indian Ocean *J. Mar. Sci.* 1:71-80.

Lyimo, T.J., Pol, A., Jetten, S.M.M. and Op den Camp H.J.M. 2008. Diversity of methanogenic archaea in a mangrove sediment and isolation of a new *Methanococcoides* strain. *FEMS Microbiol. Lett.* 291:247–253

Mackey, A.P., Hodgkinson, M. and Nardella, R. 1992. Nutrient levels and heavy metals in mangrove sediments from the Brisbane River, Australia, *Mar. Pollut. Bullet,* 24: 418-20.

Madigan, M.T., Martinko, J.M. and Parker, J. 2000. Brock Biology of Microorganism. 9th ed. Upper Saddle River, N. J.: Prentice-Hall.

Mahadevan, A. and Sridhar, R. 1986. Methods in Physiological Plant Pathology. Sivakami publication, Madras, India.

Mann, F.D. and Steinke, T.D. 1992. Biological nitrogen fixation (acetylene reduction) associated with decomposing *Avicennia marina* leaves in the Beach wood Mangroove Nature Reserve. *S. Afr. J. Bot.* 58: 533-536.

Marchand, C., Baltzerb, F., Lallier-Verge`sa, E. and Albe Ørica P. 2004. Pore water chemistry in mangrove sediments: relationship with species composition and developmental stages (French Guiana). *Mar. Geol.* 208: 361-81.

Marchand, C., Fernandez, J.M., Moreton, B., Landi, L., Lallier-Vergès, E. & Baltzer, F. 2012. The partitioning of transitional metals (Fe, Mn, Ni, Cr) in mangrove sediments downstream of a ferralitized ultramafic watershed (New Caledonia). *Chem. Geol.* 300: 70-80.

Marchand, C., Lalliet, V.E., Baltzer, F., Alberic, P., Cossa, D. and Baillif, P. 2006. Heavy metals distribution in mangrove sediments along the mobile coastline of French Guiana. *Mar. Chem.* 98: 1-17.

Margesin, R. and Schinner, F. (2001). Potential of halotolerant and halophilic microorganisms for biotechnology, *Extremophiles,* 5: 73–83.

Maria, G.L., Sridha, K.R. and Raviraja, N.S. 2005. Antimicrobial and enzyme activity of mangrove fungi of south west coast of India. *J. Agric. Technol.* 1: 67–80.

Marty, D.G. 1985. Description de quatre souches Methanogenes thermotolerants isolee de sediments marins ou intertidaux. *C. R. Acad. Sci. III.* 300: 545–548.

Maxwell, G.S. 1968. Pathogenicity and salinity tolerance of *Phytophthora* sp. isolated from *Avicennia resinifera* (ForstF.)-some initial investigations. *Tane.* 14:13–23.

Mishra, R.R., Dhal, B., Dutta, S.K., Dangar, T.K., Das, N.N. and Thatoi, H.N. 2012. Optimization and characterization of chromium(VI) reduction in saline condition by moderately halophilic Vigribacillus sp. isolated from mangrove soil of Bhitarkanika, India. *J. Hazard. Mater.* 228: 219-226.

Mishra, R.R., Prajapati, S., Das, J., Dangar, T.K., Das, N. and Thatoi, H.N. 2011 Reduction of selenite to red elemental selenium by moderately halotolerant *Bacillus megaterium* strains isolated from Bhitarkanika mangrove soil and characterization of reduced product. *Chemosphere.* 84(9): 1231–1237.

Mishra, R.R., Rath, B. and Thatoi, H.N. 2008. Water quality assessment of aquaculture ponds located in Bhitarakanika mangrove ecosystem, Orissa, India. *Turk J Fish Aqua Sci.* 8: 71-77.

Mishra, R.R., Thatoi, H.N. and Dangar, T.K. 2010. Microbial biodiversity of Bhitarkanika, Orissa- A phenotypic, genetic and proteomic characterization of the predominant bacteria. Ph.D thesis submitted to North Orissa University, Takatpur, Baripada, Orissa.

Mitra S, Pattanayak J.G. (2013). Studies On *Lingula AnaTina* (brachiopoda: Inarticulata) In Subarnarekha Estuary, Odisha With Special Reference To Habitat And Population Rec. zool. Surv. India:113(Part-3):49-53

Mitra, A., Santra, S.C. and Mukherjee, J. 2008. Distribution of actinomycetes and antagonistic behaviour with the physico-chemical characteristics of the world's largest tidal mangrove forest. *Appl. Microbiol. Biotechnol.* 80:685–695.

Mobanraju, R., Rajgopal, B.S., Daniels, L. and Natrajan, R. 1997. Isolation and characterisation of methanogenic bacteria from mangrove sediment. *J. Mar. Biotechnol.* 5:147–152.

Mohamed, M.O.S., G. Neukermans, J.G. Kairo, F. Dahdouh-Guebas, and N. Koedam. 2009. Mangrove forests in a peri-urban setting: the case of Mombasa (Kenya). Wetlands Ecology and Manage-ment17: 243–255.

Mohanta, Y.K. 2014. Isolation of cellulose degrading actinomycetes and evaluation of their cellulolytic potential. *Bioengineer. Biosci.* 2:1-5.

Morell, J.M. and Corredor, J.E. 1993. Sediment nitrogen trapping in a mangrove lagoon. *Estuar. Coast. Shelf Sci.* 37: 203-212.

Mulligan, C.N. 2009. Recent advances in the environmental applications of biosurfactants. *Curr. Opin. Colloid. Interface. Sci.* 14:372–378.

Murphy, J. and Riely, J.P. 1962. A modified single solution method for the determination of phosphate in natural waters. *Analyt. Chim. Acta.* 27: 31–36.

Murashige, T. & Skoog, F. 1962. A revised medium for rapid growthand bio-assays with tobacco tissue cultures. *Physiol. Plant.* 15(3): 473---497.

Muyzer, G. and Stams, A.J.M. 2008. The ecology and biotechnology of sulphate-reducing bacteria. *Nat. Rev. Microbiol.* 6: 441–454.

Nakano Y. and Asada, K. 1981. Hydrogen peroxide is scavenged by ascorbate specific peroxidise in spinach chloroplasts. *Plant Cell Physiol.* 22: 867-880.

Narayanswami, V. And H.G.Carter (1922): Systematic list of the plants of Barkuda. Mem. Asiat. Soc. Beng. 7(4): 289-319.

Nautiyal, C.S. 1999. An efficient microbiological growth medium for screening phosphate solubilizing microorganism. *FEMS Microbiol. Lett.* 170: 265–270.

Naylor, R.L., Goldburg, R.J., Mooney, H., Beveridge, M., Clay, J., Folke, C., Kautsky, N., Lubcheno, J., Primavera, J., Williams, M., 2000. Nature's subsidies to shrimp and salmon farming. Science 282, 883–884

Nedwell, D.B., Blackburn, T.H. and Wiebe, W.J. 1994. Dynamic nature of the turnover of organic carbon, nitrogen and sulphur in the sediments of a Jamaican mangrove forest. *Mar. Ecol. Prog. Ser.* 110: 223-231.

Nepple, B.B., Kessi, J. and Bachofen, R. 2000. Chromate reduction by *Rhodobacter sphaeroides. J. Ind. Microbiol. Biotechnol.* 25: 198–203.

Newell, S.Y. 1996. Established and potential impacts of eukaryotic mycelia decomposers in marine/estuarine ecotones. *J. Experiment. Mar. Biol. Ecol.* 200: 187-206.

Nickerson, N.H.S. and Thiodeau, F.R. 1985. Association between pore water sulphide concentration and the distribution of mangroves. *Biogeochem.* 1: 183-192.

Nissenbaum, A. and Swaine, D.J. 1976. Organic matter-metal interactions in recent sediments the role of humic substances. *Geochim. Cosmochim. Acta.* 40: 809-816.

Nriagu, J.O. 1996. Toxic metal Pollution in Africa. *Science.* 223: 272.

Odum, W.E. and Heald, E.J. 1972. Trophic analyses of an estuarine mangrove community. *Bull. Mar. Sci.* 22: 671-738.

Odum, W.E. and Heald, E.J. 1975a. Mangrove forests and aquatic productivity.In: Hasler AD (ed) Coupling of land and water systems. Ecological studies series. Springer, Berlin, pp 129–136

Odum, W.E. and Heald, E.J. 1975b. The detritus-based food web of an estuarine mangrove community. In: Ronin LT (ed) Estuarineresearch. Academic, New York, pp 265–286.

Ong JE (1995) The ecology of mangrove conservation and management. Hydrobiologia 295: 343–351.

Oremland R.S., Marsh, L.M. and Polcin, S. 1982. Methane production and simultaneous sulfate reduction in anoxic salt marsh sediments. *Nature.* 296:143–145

Pal, A.K. and Purkayastha, R.P. 1992. New parasitic fungi from Indian mangrove. *J Mycopathol. Res.* 30:173–176.

Panchnadikar, V.V. 1993. Studies of iron bacteria from mangrove ecosystem in Goa and Konkan. *Int. J. Environ. Stud.* 45(1): 17–21.

Panda, S.P., Subudhi, H. and Patra, H.K. 2013. Mangrove forests of River estuaries of Odisha, India. *Int. J. Biodivers Conserva.* 5(8): 446-454.

Pandav, B. 1997. Birds of Bhitarkanika mangrove, Eastern India. *Forkt*.12: 7-17.

Paul, E.A. and Clark, F.E. 1988. Soil microbiology and biochemistry. San Diego, CA: Academic Press.

Pegg, K.G., Gillespie, N.C. and Forsberg, L.I. 1980. *Phytophthora* spp. associated with mangrove death in central coastal Queensland. *Aust. Plant. Pathol.* 9: 6–7.

Pekey H. 2006. Heavy Metals Pollution Assessment in Sediments of the Izmit Bay, Turkey. *Environ Monit. Ass.* 123: 219-31.

Pelegri, S.P. and Twilley, R.R. 1998. Heterotrophic nitrogen fixation (acetylene reduction) during leaf litter decomposition of two mangrove species from South Florida, USA. *Mar. Biol.* 131(1): 53–61.

Poch, G.K. and Gloer, J.B. 1991. Auranticins A and B: two depsidones from a mangrove isolate of the fungus *Preussia aurantiaca*. *J. Nat. Prod.* 54: 213–217.

Polizeli, M.L.T.M., Rizzatti, A.C.S., Monti, R., Terenzi, H.F., Jorge, J.A. and Amorim, D.S. 2005. Xylanases form fungi: properties and Industrial applications. *Appl. Microbiol. Biotechnol.* 67: 577–591.

Prabhakarann, N. and Gupta, R. 1990. Activity of soil fungi of Mangalvan, the mangrove ecosystem of Cochin backwater. *Fish. Technol.* 27: 157–159.

Prasad, M.B.K. 2005. Nutrient dynamics in Pichavaram mangroves, southeast coast of India. Ph.D.Thesis. Jawaharlal Nehru University, New Delhi, India

Primavera, J.H., 2005. Mangroves, fishponds, and the quest for sustainability. Science 310, 57–59.

Purvaja, R., Ramesh, R. and Frenzel P. 2004. Plant-mediated methane emission from an Indian mangrove. *Glob. Cha. Biol.* 10: 1825-1834.

Qasim, S.Z. (1998). Mangroves, In :Glimpses of the Indian Ocean, (University Press, Hyderabad), pp. 123-129

Raghukumar, C., Muraleedharan, U., Gaud, V.R. and Mishra, R. 2004. Xylanases of marine fungi of potential use of bioleaching of paper pulp. *J. Ind. Microbiol. Biotechnol.* 31: 433–441.

Raghukumar, S., Sathe-Pathak, V., Sharma, S. and Raghukumar, C. 1995. *Thraustochytrid* and fungal component of marine detritus. Field studies on decomposition of leaves of the mangrove *Rhizophora apiculata*. *Aquat. Microb. Ecol.* 9: 117–125.

Rahman, M.M., Ullah, M.R., Lan, M., Sumantyo, J.T., Kuze, H. and Tateishi, R. 2013. Comparison of Landsat image classification methods for detecting mangroveforests in Sundarbans. *Int. J. Remote Sens.* 34: 1041-1056.

Rajkumar, J., Swarnakumar, N.S., Sivakumar, K., Thangardajou, T. and Kannan L. 2012. Actinobacterial diversity of mangrove environment of the Bhitherkanika mangroves, east coast of Orissa, India. *Int. J. Sci. Res. Public.* 2: 1-6.

Ramachandran, S. and Venugopalan, V.K. 1987. Nitrogen fixation by blue green algae in porto Novo Marine environment. *J. Mar. Biol. Ass. Ind.* 29 (1-2): 337-343.

Ramamurthy, T., Raju, R.M. and Natarajan, R. 1990. Distribution and ecology of methanogenic bacteria in mangrove sediments of Pichavaram, east coast of India. *Ind. J. Mar. Sci.* 19: 269–273.

Ramanathan, A.L., Singh, G., Majumdar, J., Samal, A.C., Chauhan, R., Ranjan, R.K., Rajkumar, K. and Santra, S.C. 2008. A study of microbial diversity and its interaction with nutrients in the sediments of Sundarban mangroves. *Ind. J. Marine Sci.* 37(2): 159-165.

Ramsay, M.A., Swannell, R.P.J., Shipton, W.A., Duke, N.C. and Hill, R.T. 2000. Effect of bioremediation community in oiled mangrove sediments. *Mar. Pollut. Bull.* 41: 413–419.

Raven, J.A. and Scrimgeour, C.M. 1997. The influence of anoxia on plants of saline habitats with special reference to the sulphur cycle. *Ann. Bot.*79: 79-86.

Ravichelvan R., Ramu S. and Anandraj, T. 2015. Seasonal variations of water quality parameters in south east Coastal waters of tamil nadu, india. *nt. J. Modn. Res. Revs.* 3(10): 826-829.

Ravikumar, D.R. and Vittal, B.P.R. 1996. Fungal diversity on decomposing biomass of mangrove plant Rhizophora in Pichavaram estuary, east coast of India. *Indian J. Mar. Sci.* 25: 142-144.

Ravikumar, S. 1995. Nitrogen fixing *Azotobacters* from the mangrove habitat and their utility as biofertilizers. Ph.D. thesis, Annamalai University, India, pp.202.

Ravikumar, S., Fredimoses, M. and Gokulakrishnan, R. 2011. Biodiversity of actinomycetes in Manakkudi mangrove ecosystem, Southwest coast of India. *Ann. Biol. Res.* 2(1):76–82.

Ravikumar, S., Inbaneson, S.J., Uthiraselvam, M., Ramu, A. and Banerjee, M.B. 2010. Diversity of endophytic actinomycetes from Karangkadu mangrove ecosystem and its antibacterial potential against bacterial pathogens. *J. Pharm. Res.* 4(1): 294–296.

Ray, A.K., Bairagi, A., Ghosh, S., Sen, S.K. (2007). Optimization of fermentation conditions for cellulase production by *Bacillus subtilis* CY5 and *Bacillus circulans* TP3 isolated from fish gut. Acta Ichthyologica et Piscatoria. 37(1):47-53.

Reef, R., Feller, I.C. and Lovelock, C.E. 2010. Nutrition of mangroves. *Tree Physiol.* 30(9): 1148–1160.

Rivera-Monroy, V.H., Day, W.J., Twilley, R.R., Vera-Herrera, F., Coronado-Molina, C. (1995a). Flux of nitrogen and sediment in a fringe mangrove forest in Terminos lagoon, Mexico. Estuar. Cost. Shelf Sci. 40:139-160.

Robertson, A. I. 1986. Leaf-burying crabs: Their influence on energy flow and export from mixed mangrove forests (*Rhizophora* spp.) in northeastern Australia, *J. Exp. Mar. Biol. Ecol.* 102: 237–248.

Robertson, A. I., Alongi, D. M. and Boto, K. G. 1992. Food chains and carbon fluxes, In Robertson, A.I., and Alongi eds. pp. 293–326.

Rodriguez, H. and Fraga, R. 1999. Phosphate solubilizing bacteria and their role in plant growth promotion. *Biotech. Adv.*17: 319-339.

Rojas, A., Holguin, G., Glick, B.R. and Bashan, Y. 2001. Synergism between *Phyllobacterium* sp.(N2-fixer) and *Bacillus licheniformis* (P-solubilizer), both from a semi arid mangrove rhizosphere. *FEMS Microbiol Ecol*. 35: 181–191.

Rovira, M., Gimenez, J., Martınez, M., Martınez-Llado, X., DePablo, J., Marti, V. and Duro, L. 2008. Sorption of selenium(IV) and selenium(VI) onto natural iron oxides: goethite and hematite. *J. Hazard. Mater*. 150: 279–284.

Roy, S., Hens, D., Biswas, D., Biswas, D. and Kumar, R. 2002. Survey of petroleum - degrading bacteria in coastal waters of Sunderban Biosphere Reserve. *World J. Microbiol. Biotech*. 18: 575-581.

Sabu, A. 2003. Sources, properties and applications of microbial therapeutic enzymes. *Ind. J. Biotechnol*. 2(3): 334–341.

Sadaba, R.B., Vrijmoed, L.L.P., Jones, E.B.G. and Hodgkiss, I.J. 1995. Observations on vertical distribution of fungi associated with standing senescent *Acanthus ilicifolius* stems at Mai Po mangrove, Hong Kong. *Hydrobiol*. 295: 119–126.

Sadhu, S., Saha, P., Sen, S.K., Mayilraj, S. and Maiti, T.K. 2013. Production, purification and characterization of a novel thermotolerant endoglucanase (CMCase) from *Bacillus* strain isolated from cow dung. Springerplus 2: 10. http://www.springerplus.com/content/2/1/10.

Saenger, P., Hegerl, E.J. and Davie, J.D.S. (Eds.), 1983. Global Status of Mangrove Ecosystems. The Environmentalist 3 (Supplement):1-88.

Sahoo, K. and Dhal, N.K. 2009. Potential microbial diversity in mangrove ecosystem: A review. *Ind. J. Mar. Sci*. 38(2): 249-256.

Sahu, M.K., Sivakumar, K. and Kannan, L. 2005. Isolation of actinomycetes from different samples of the Vellar estuary, southeast coast of India. *Pollut. Res*. 24: 45–48.

Sahu, S.C., Sahoo, K., Kumar, P. and Dhal, N.K. 2013. Floral and microbial dynamics in relation to the Physico-chemical constituents of the Devi-estuary of Odisha Coast of the Bay of Bengal, India. *Ind. J. Geo-Mar. Sci*. 42(1): 90-96.

Saimmai, A., Sobhon, V. and Maneerat, S. 2011. Production of biosurfactants from a new and promising strain of *Leucobacter komagatae* 183. *Ann. Microbiol*. 62(1): 391–402.

Samira, M., Mohammed, M. and GholamReza, G. 2011. Carboxymethyl cellulase and Filter-paperase activity of new strains isolated from Persian gulf. *Microbiol. J*. 1: 8–16.

Santhi, V.S. and Jebakumar, S.R.D. 2011. Phylogenetic analysis and antimicrobial activities of *Streptomyces* isolates from mangrove sediment. *J. Basic. Microbiol*. 51: 71–79.

Santos, H.F., Cury J.C., Carmol, F.L., Santos, A.L., Tiedje, J., Elsas, J.D., Rosado, S.A. and Peyote, R.S. 2011. Mangrove bacterial diversity and the impact of oil contamination revealed by pyrosequencing: bacterial proxies for oil pollution. *PLoS ONE* 6(3): e16943.

Sarangi, R.K., Kathiresan, K. and Subramanian, A.N. 2002. Metal concentrations in five mongrove species of the Bitharkanika, Orissa, east coast of India. *Ind. J. Mar. Sci.* 31(3): 251-253.

Sarma, V.V. and Hyde, K.D. 2001. A review on frequently occurring fungi in mangroves. *Fung. Diver.* 8: 1–34.

Sarma, V.V. and Vittal, B.P.R. 2001. Biodiversity of manglicolous fungi on selected plants in the Godavari and Krishna deltas, East Coast of India. *Fung. Div.* 6: 115–130.

Sasirekha, B., Bedashree, T. & Champa, K. L. 2012. Optimizationand partial purification of extracellular phytase from *Pseu-domonas aeruginosa p6. Europ. J. Experiment. Biol.* 2(1): 95---104.

Saxena, D., Loka-Bharathi, P.A. and Chandramohan, D. 1988. Sulfate reducing bacteria from mangrove swamps of Goa, central west coast of India. *Ind. J. Mar. Sci.* 17: 153-157.

Selander, R. K., Caugant, D. A., Ochman, H., Musser, J. M., Gilmour, M. N. & Whittam, T. S. 1986. Methods of multilocus enzyme electrophoresis for bacterial population genetics and systematics . Appl. Environ. Microbiol. 51: 8 37-884.

Selvam V, Gnanappazham L, Navamuniyammal M, Ravichandran KK, Karunakaran VM (2002) Atlas of mangrove wetlands of India (Part-I) M.S. Swaminathan Research Foundation, Chennai, Tamilnadu, p. 100.

Selvam, V. 2003. Environmental classification of man-grove wetlands of India. Current Science 84: 757-765.

Selvam, V. and Karunagaran, V. M. 2004. Coastal wetlands: Mangrove conservation and management. Orientation guide 1. Ecology and biology of mangroves. M.S. Swaminathan Research Foundation, Chennai.

Selvam, V., Eganathan, P., Karunagaran, V.M., Ravishankar, T., Ramasubramanian, R. 2004. Mangrove Plants of Tamil Nadu. M.S.Swaminathan Research Foundation, Chennai, India.

Sen, N. and Naskar, K. 2003. Algal flora of Sundarbans Mangal. Daya books publication.

Sengupta, A. and Choudhury, S. 1990. Halotolerant *rhizobium* strains mangrove swamps of the Gangas River Delta. *Ind. J. Microbiol.* 30(4): 483-484.

Sengupta. A. and Chaudhuri, S. 2002. Arbuscular mycorrhizal relations of mangrove plant community at the Ganges river estuary in India. *Mycorrhiza.* 12: 169–174.

Shaiful, A.A.A., Abdul Manan, D.M., Ramli, M.R. and Veerasamy, R. 1986. Ammonification and nitrification in wet mangrove soils. *Mal. J. Sci.* 8:47-56.

Shanmugam, S. and Mody, K.H. 2000. Heparonid active sulphated polysaccharides from marine algae as potential blood coagulant agents. *Curr. Sci.* 79: 1672–1683

Shearer, C.A., Descals, E., Kohlmeyer, B., Kohlmeyer, J., Marvanová, L., Padgett, D., Porter, D., Raja, H.A., Schmit, J.P. and Thorton, H.A.et al. 2007. Fuangal diversity in aquatic habitats. *Biodiver. Conserv.* 16: 49–67.

Sherman, R.E., Fahey, T.J. and Howarth, R.W. 1998. Soil-plant interactions in a neotropical mangrove forest: Iron, phosphorus and sulfur dynamics. *Oecol.* 115: 553-563.

Shoreit, A.A.M., EL- Kady, I.A. and Sayed, W.F. 1994. Isolation and identification of purple non sulphur bacteria of mangal and non-mangal vegetation of red sea coast, Egypt. *Limnol.* 24: 177–183.

Shriadah, M.M.A. 1999. Heavy metals in mangrove sediments of the United Arab Emirates Shoreline (Arabian Gulf). *Water Air Soil. Poll.* 116: 523-34.

Siddique, T., Zhang, Y., Okeke, B.C. and Frankenberger Jr., W.T. 2006. Characterization of sediment bacteria involved in selenium reduction. *Biores. Technol.* 97: 1041–1049.

Silva C.A.R., Lacerda, L.D. and Rezende, C.E. 1990. Heavy metal reservoirs in a red mangrove forest, *Biotropic.* 22. 339-345.

Sivakumar, K. 2001. Actinomycetes of an Indian mangrove (Pichavaram). Environment. PhD thesis, Annamalai University.

Sivakumar, K., Sahu, M.K. and Kathiresan, K. 2005. An antibiotic producing marine Streptomyces from the Pichavaram mangrove environment. *J. Ann. Univ. Part-B* XLI: 9–18.

Song, X.H., Liu, X. and Lin, Y.C. 2004. Metabolites of mangrove fungus No. K23 and interaction of carboline with DNA. *J. Trop. Oceangr.* 23(3): 66–71.

Spalding, M., Kainuma, M. and Collins, L., 2010. World atlas of mangroves. Earthscan, London.

Staley, J.T. and Gosink, J.J. 1999. Poles apart: biodiversity and biogeography of sea ice bacteria. *Annu. Rev. Microbiol.* 53: 189-215.

State of Forest Report, 2003. Forest Survey of India.

Steinbüchel, A. and Fuchtenbusch, B. 1998. Bacteria and other biological systems for polyester production. *Trend. Biotechnol.* 16: 419–427.

Steinke, T.D. and Ward, C.J. 1987. Degradation of mangrove leaf litter in the St Lucia Estuary as influenced by season and exposure. *South Afric. J. Bot.* 53: 323-328.

Sudha, V. 1981. Studies (a) on halophilic bacteria associated with mangrove sediment and a biovalve mollusc *Anadara rhombea* (Born) (Arcidae) and (b) on L-asparaginase (Anti Leukamicagent) from an extremely halophilic bacterium. PhD Thesis. Annamalai University, Parangipettai, India.

Sundararaj, V., Dhevendran, K., Chandramohan, D. and Krishnamurthy, K. 1974. Bacteria and primary production. *Ind. J. Mar. Sci.* 3: 139-141.

Suryanarayanan, T.S., Kumaresan, V. and Johnson, J.A. 1998. Foliar fungal endophytes from two species of the mangrove Rhizophora. *Can. J. Microbiol.* 44: 1003–1006.

Tabao, N.C. and Moasalud, R.G. 2010. Characterisation and identification of high cellulose-producing bacterial strains from philipine mangroves. *Philipp. J. Syst. Biol.* IV.

Tabatabi, M.A. and Bremner, J.M. 1969. Use of P-nitrophynyl phosphate for assay of soil phosphatase activity. *Soil Biol. Biochem.* 1: 301–307.

Taketani, G.R., Yoshiura, A.C., Dias, F.C.A., Andreote, D.F. and Tsai, M.S. 2010. Diversity and identification of methanogenic archaea and sulphate-reducing bacteria in sediments from a pristine tropical mangrove. *Ant. van Leeu. Hoek.* 97: 401–411

Tam, N.F. and Wong, Y.S. 1995. Spatial and temporal variation of heavy metal contamination in sediments of a mangrove swamp in Hong Kong. *Mar. Poll. Bullet.* 31: 254-61.

Tam, N.F.Y. & Wong, W.S. 2000. Spatial variation of heavy metals in surface sediments of Hong Kong mangrove swamps. *Environ. Pollut.* 110: 195-205.

Tao, L., Zhang, J.Y., Liang, Y.J., Chen, L.M., Zhen, L.C., Wang, F., Mi, Y., She, Z.G., To, K.K.W., Lin, Y.C. and Fu, L.W. 2010. Anticancer effect and structure activity analysis of marine products isolated from metabolites of mangrove fungi in the South China Sea. *Mar. Drug.* 8:1094–1105.

Taylor, L.E., Henrissat, B., Coutinho, P.M., Ekborg, N.A., Hutcheson, S.W. and Weiner, R.M. 2006. Complete cellulase system in the marine bacterium *Saccharophagus degradans* strain 2-40T. *J. Bacteriol.* 188: 3849-3861.

Thatoi, H.N. and Biswal. A.K. 2008. Mangroves of Odisha coast: floral diversity and conservation status. Special habitats and threatened plants of India. ENVIS Wildlife and protected area. 11: 201-207.

Thatoi, H.N., Behera, B.C., Danger, T.K. and Mishra, R.R. 2012. Microbial biodiversity in mangrove soil of Bhitarakanika, Odisha, India. *Int. J. Environ. Biol.* 2(2): 50-8.

Tolan, J.S. and Foody, B. 1999. Cellulase from submerged fermentation. In: Tsao GT, editor. Advances in biochemical engineering biotechnology. Recent progress in bioconversion of lignocellulosics, vol. 65. Berlin: Springer-Verlag.p. 41–67.

Toledo, G., Bashan, Y. and Soeldner, A. 1995a. Cyanobacteria and black Mangrooves in North Western Mexico. Colonization and diurnal and seasonal nitrogen fixation on aerial roots. *Can. J. Microbiol.* 41: 999-1011.

Torsvik, V., Goksoyr, J., Daae, F.L. 1990. High diversity in DNA of soil bacteria. *Appl. Environ. Microbiol.* 56: 782-787.

Twilley, R.R. and Chen, R. 1998. A water budget and hydrology model of a basin mangrove forest in Rookery Bay, Florida. *Mar. Fresh water Res.* 49: 309–323.

Uchino, F., Hambali, G.G. and Yatazawa, M. 1984. Nitrogen-fixing bacteria from warty lenti cellate bark of a mangrove tree, *Bruguiera gymnorrhiz* (aL .) *Lamk.* *Appl. Environ. Microbiol.* 47: 44-48.

Urakawa, H., Kita-Tsukamoto, K. and Ohwada, K. 1999. Microbial diversity in marine sediments from Sagami Bay and Tokyo Bay, Japan, as determined by 16S rRNA gene analysis. *Microbiol.* 145: 3305-3315.

Varon-Lopez, M., Dias, A.C.F., Fasanella, C.C., Durrer, A., Melo, I.S., Kuramae, E. E. and Andreote, F. D. 2014. Sulphur-oxidizing and sulphate-reducing communities in Brazilian mangrove sediments. *Environ. Microbiol.* 16: 845 –855.

Vassileva, M., Azcon, R., Barea Miguel, J. and Vassilev, N. 1998. Application of an encapsulated filamentous fungus in solubilisation of inorganic phosphate. *J. Biotechnol.* 63(1): 67-72.

Vazquez, P., Holguin, G., Puerte, M.E., Lopez-Cortes, A. and Bashan, Y. 2000. Phosphate-solubillising micro organisms associated with the rhizosphere of mangroves in a semiarid Coastal lagoon. *Biol. Fertil. Soils.* 30: 460-468.

Venkateswara, Sarma, V., Hyde, K.D. & Vittal, B.P. 2001. Frequency of occurrence of mangrove fungi from the east coast of India. *Hydrobiol.* 455: 41-53.

Venkateswaran, K. and Natarajan, R. 1983. Seasonal distribution of inorganic phosphate solubilising bacteria and phosphatase producing bacteria in Porto Novo waters. *Ind. J. Mar. Sci.* 12(4): 213-217.

Ventosa, A. and Nieto, J.J. 1995. Biotechnological applications and potentialities of halophilic microorganisms. *World J. Microbiol. Biotechnol.* 11: 85-94.

Venugopal, M. and Saramma, A.V. 2006. Characterization of alkaline protease from *Vibrio fluvialis* strain VM10 isolated from a mangrove sediment sample and its application as a laundry detergent additive. *Proc. Biochem.* 41:1239–1243.

Vethanayagam, R.R. 1991. Purple photosynthetic bacteria from a tropical mangrove environment. *Mar. Biol.* 110:161-163.

Vethanayagam, R.R. and Krishnamurthy, K. 1995. Studies on anoxygenic photosynthetic bacteria *Rhodopseudomonas* sp. from the tropical mangrove environment. *Ind. J. Mar. Sci.* 24:19-23.

Vreeland, R.H. 1987. Mechanisms of halotolerance in microorganisms. *Crit. Rev. Microbiol.* 14: 311–356.

Vrijmoed, L.L.P., Iones, E.B.G. and Alias, S.A. 1996. Preliminary observations on marine fungi and mangrove fungi from Hainan Island in South China Sea. *Asian J. Tropic. Biol.* 2: 31-36.

Walters, B., Rönnbäck, P., Kovacs, J., Crona, B., Hussain, S. et al., 2008. Ethnobiology, socio-economics and management of mangrove forests: a review. Aquatic Bot 89(2), 220–236.

Wang, W., Liu, J., Chen, G., Zhang, Y. and Gao, P. 2003. Function of a low molecular weight peptide from *Trichoderma pseudokoningii* S38 during cellulose biodegradation. *Curr. Microbiol.* 46: 371-379.

Wiwin, R. 2010. Identification of Streptomyces sp-MWS1 producing antibacterial compounds. *Indo. J. Trop. Infect. Dis.* 1(2): 80–85.

Wu, R.Y. 1993. Studies on the microbial ecology of the Tansui Estuary. *Bot. Bull. Acad. Sin.* 34:13–30.

Xie, X.C., Mei, W.L., Zhao, Y.X., Hong, K. and Dai, H.F. 2006. A new degraded sesquiterpene from marine actinomycete *Streptomyces* sp.0616208. *Chin. Chem, Lett.* 17:1463–1465.

Xin, Li., Ryuichiro, K. and Kokki S. 2003. Studies on hypersaline– tolerant white–rot fungi IV: effects of Mn and NH4 on manganese peroxidase production and

Roly R–478 decolorization by the marine isolate *phlebia* sp. MG–60 under saline conditions. J. *Wood Sci.* 49: 355–360.

Xin, Li., Ryuichiro, K. and Kokki, S. 2002. Biodegradation of sugarcane bagasse with marine fugus *phlebia* sp. MG–60. *J. Wood. Sci.* 48:159–162.

Xu, M.J., Gessner, G., Groth, I., Lange, C., Christner, A., Bruhn, T., Deng, Z.W., Li, X., Heinemann, S.H., Grabley, S., Bringmann, G., Sattler, I. and Lin, W.H. 2007. Shearing D–K, new indole triterpenoids from an endophytic *Penicillium* sp. (strain HKI0459) with blocking activity on large-conductance calcium-activated potassium channels. *Tetrahed.* 63:435–444.

Yakimov, M.M., Abraham, W.R., Meyer, H., Laura, G. and Golyshin, P.N. 1999. Structural characterisation of lichenycin. A component by fast atom bombardment tandem mass spectrometry. *Biochem. Biophys. Acta.* 1438: 230–280.

Yin, L.J., Huang, P.S. and Lin, H.H. 2010. Isolation of cellulase-producing bacteria and characterization of the cellulase from the isolated bacterium *Cellulomonas* sp. YJ5. *J. Agric. Food. Chem.* 58: 9833–9837.

You, J.L., MaoW-Zhou, S.N., Wang, J., Lin Y.C. and Wu, S.Y. (2006) Fermentation conditions and characterization of endophytic fungus #732 producing novel enniatin G fromSouth China Sea. *Act. Sci. Nat.* 45(4): 75–78.

Youssef, T. and Saenger, P. 1998. Photosynthetic gas exchange and accumulation of phytotoxins in mangrove seedlings in response to soil physico-chemical characteristics associated with water logging. *Tree Physiol.* 18 (5): 317-324.

Yu, K.S., Wong, A.H., Yau, K.W., Wong, Y.S. and Tam, N.F. 2005. Natural attenuation, bio stimulation and bioaugmentation on biodegradation of polycyclic aromatic hydrocarbons (PAHs) in mangrove sediments. *Mar. Pollut. Bull.* 51: 1071–1077.

Zahran, H. H., M. S. Ahmed, and E. A. Afkar.1995. Isolation and characterization of nitrogen-fixing moderate halophilic bacteria from saline soils of Egypt. J. Basic Microbiol. 35:269–275.

Zhang, Y., Amrhein, C. and Frankenberger Jr., W.T. 2005. Effect of arsenate and molybdate on removal of selenate from an aqueous solution by zero-valent iron. *Sci. Tot. Environ.* 350: 1–11.

Zhang, Y.H.P. and Lynd, L.R. 2004. A functionally-based model for hydrolysis of cellulose by fungal cellulase. *Biotechnol. Bioeng.* 94: 888-98.

Zuberr, D.A. and Silver, W.S. 1978. Biological dinitrogen fixation (Acetylene reduction) associated with florida mangroves. *Appl. Environ. Microbiol.* 35: 567-575.

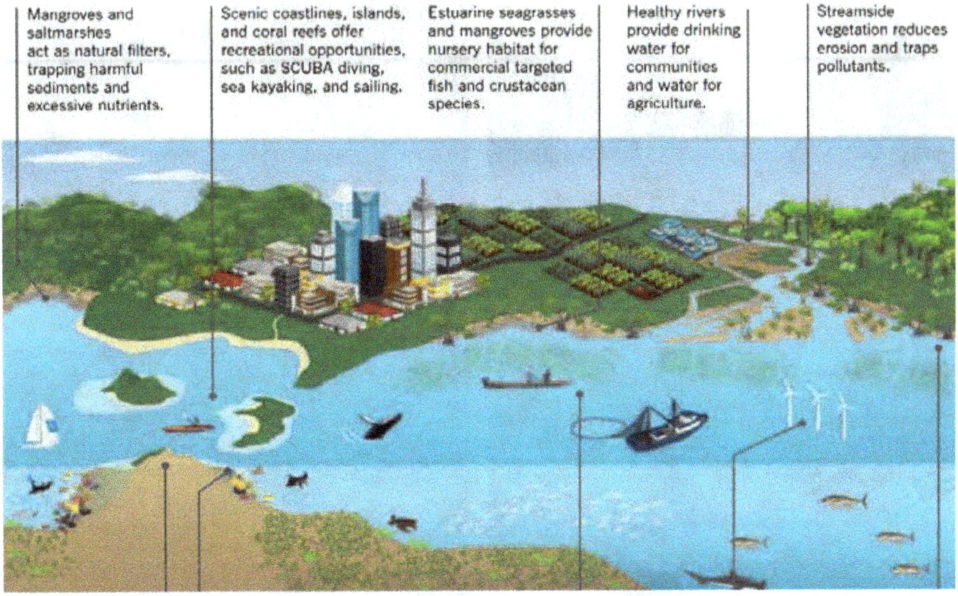
Mangroves and saltmarshes act as natural filters, trapping harmful sediments and excessive nutrients.
Scenic coastlines, islands, and coral reefs offer recreational opportunities, such as SCUBA diving, sea kayaking, and sailing.
Estuarine seagrasses and mangroves provide nursery habitat for commercial targeted fish and crustacean species.
Healthy rivers provide drinking water for communities and water for agriculture.
Streamside vegetation reduces erosion and traps pollutants.

Fig. 1 Some important uses of mangrove (p. 3)

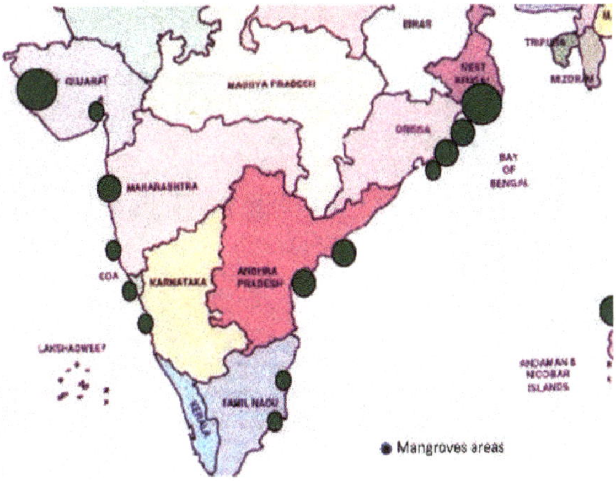
GUJARAT
MADHYA PRADESH
WEST BENGAL
ODISHA
BAY OF BENGAL
MAHARASHTRA
GOA
KARNATAKA
ANDHRA PRADESH
LAKSHADWEEP
TAMIL NADU
KERALA
ANDAMAN & NICOBAR ISLANDS
Mangroves areas

Fig. 2 Distribution of mangroves in India (p. 5)

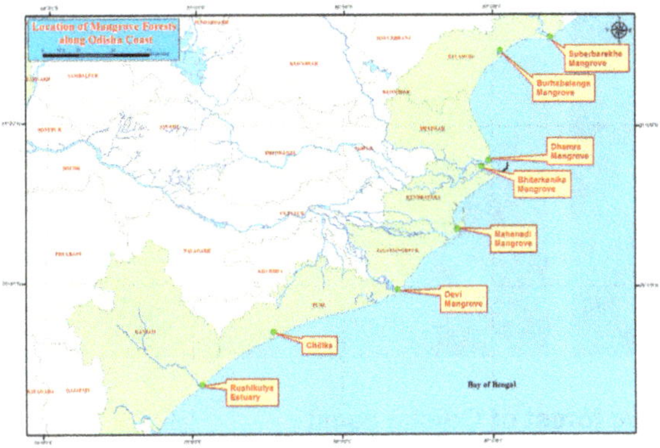
Location of Mangrove Forests along Odisha Coast
Subarnarekha Mangrove
Budhabalanga Mangrove
Dhamra Mangrove
Bhitarkanika Mangrove
Mahanadi Mangrove
Devi Mangrove
Chilika
Rushikulya Estuary
Bay of Bengal

Fig.3 Distribution of Mangrove in Odisha (p. 7)

Fig.4 Mangrove forest of Odisha coast

(p. 9)

Fig.5 Map of Mangroves of Bhitarkanika Odisha (p. 11)

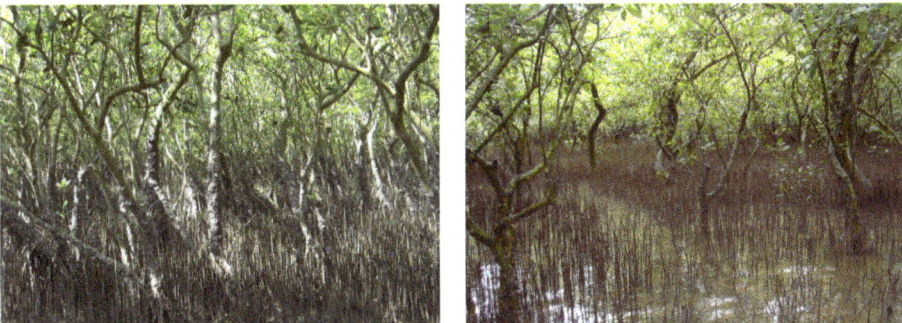

Fig.6: Bhitarkanika mangrove forest (p. 12)

Xylocarpus granatum

Kandelia candel

Lumnitzera racemosa

Sonneratia grifithii

Sesuvium portulacastrum

Pnematophores of *officinalis*

Fig. 16. Mangrove species found in Odisha coast (Thatoi & Biswal, 2008) (p. 37)

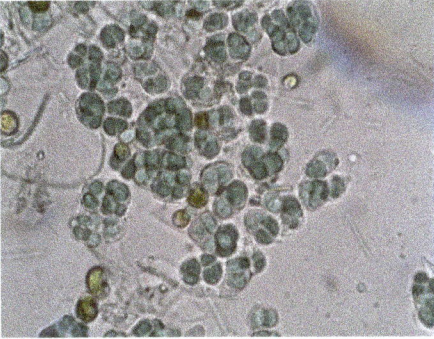

(a) *Oscillatoria* sp.

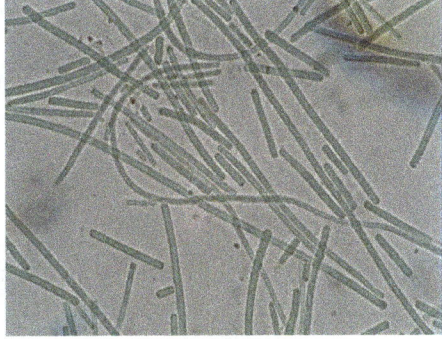

(b) *Gloeocapsa* sp.

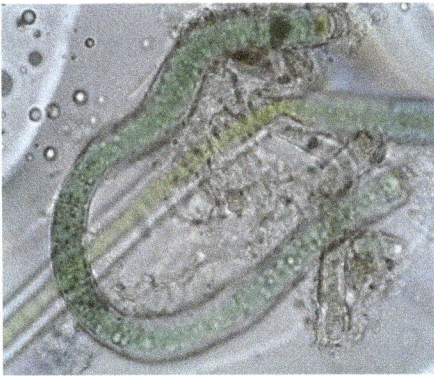

(c) *Anabaena* sp.

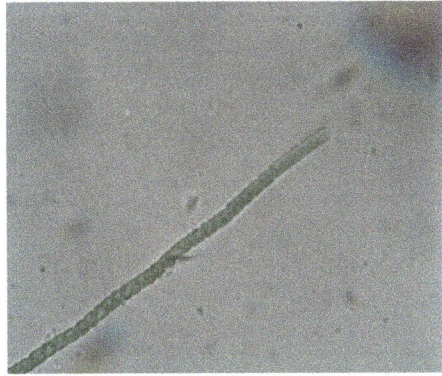

(d) *Phormidium* sp.

Figure 19: Phase contrast photograph of some algal species identified form Bhitarkanika mangrove ecosystem of Odisha. (a) *Oscillatoria* sp., **(b)** *Gloeocapsa* sp. **(c)** *Anabaena* sp. and **(d)** *Phormidium* sp. (p. 52-53)

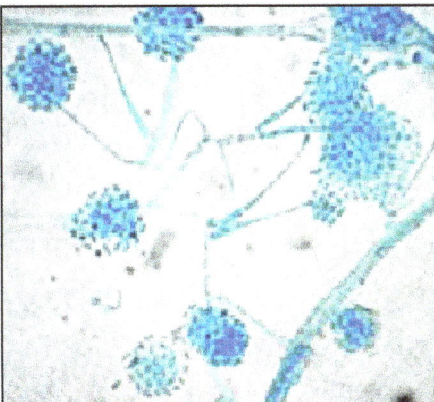

1. *Acremonium byssoides*

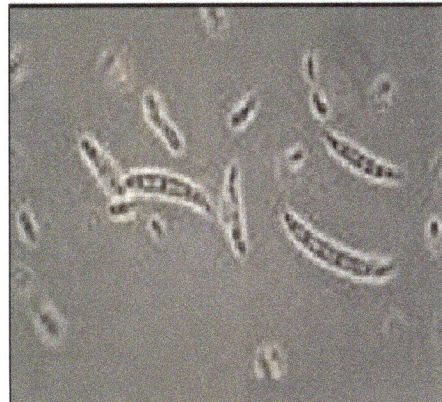

2. *Alternaria alternata*

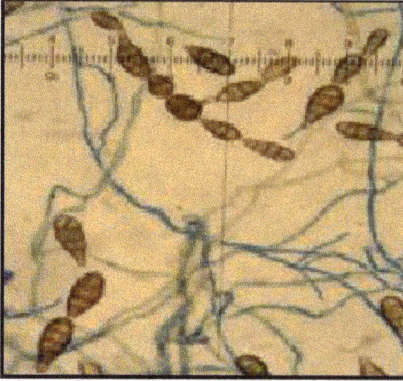

3. *Aspergillus glavus*

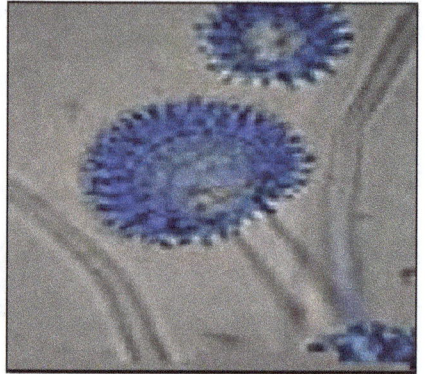

4. *Aspergillus niger*

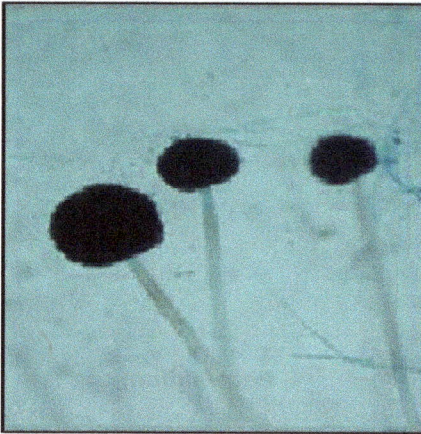

5. *Aspergillus oryzae*

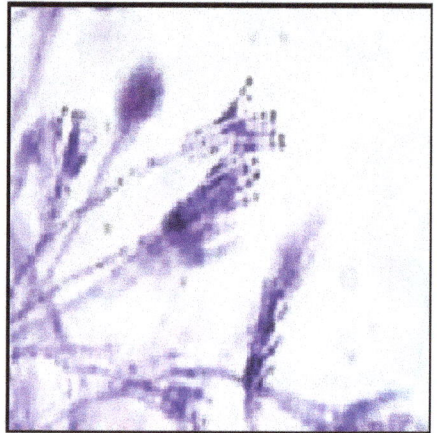

6. *Choenophora cucrbitarum*

7. *Cladosporium oxysporum*

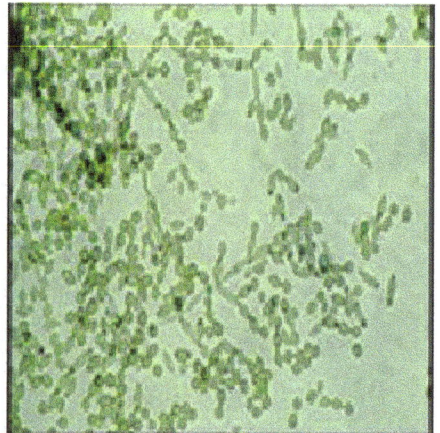

8. *Curvularia lunata*

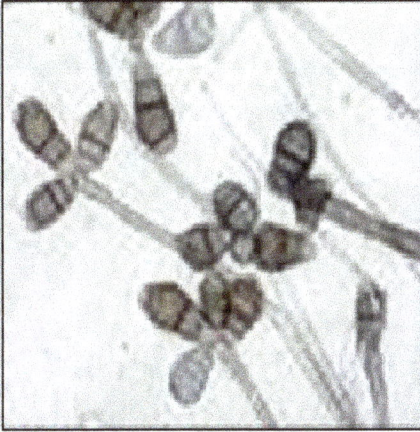

9. *Drechslera hawaiensis*

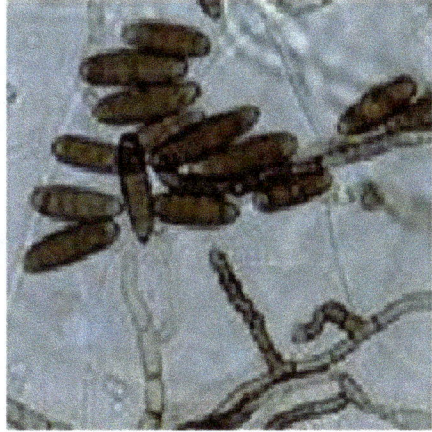

10. *Fusarium oxysporum*

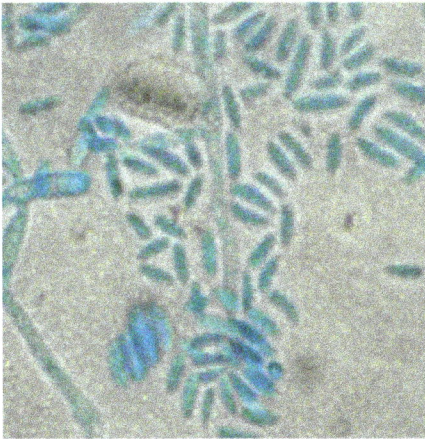

11. *Fusarium moniliforme*

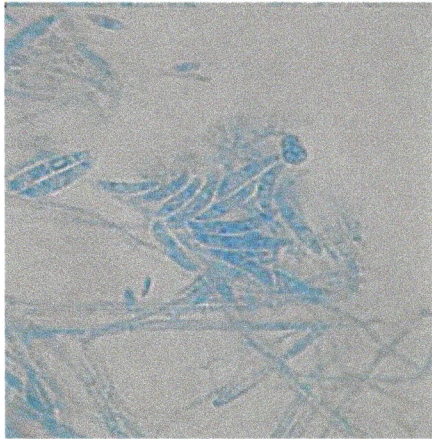

12. *Fusarium solani*

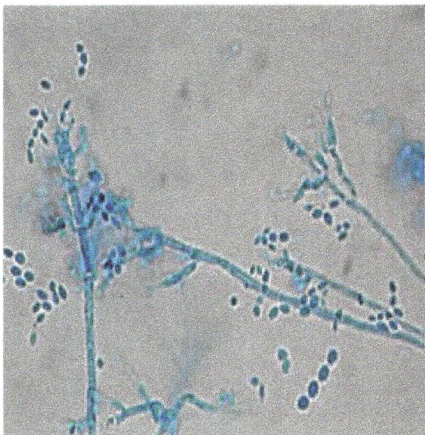

13. *Paecilomyces lilacinous*

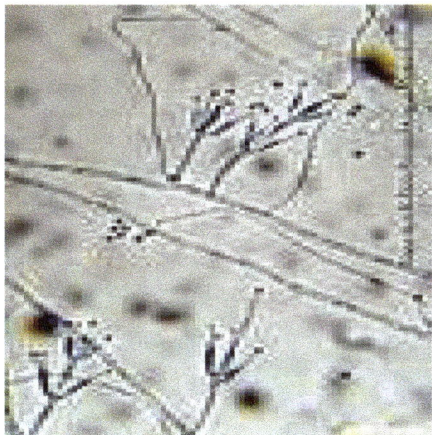

14. *Paecilomyces varioti*

15. *Penicillium digitatum*

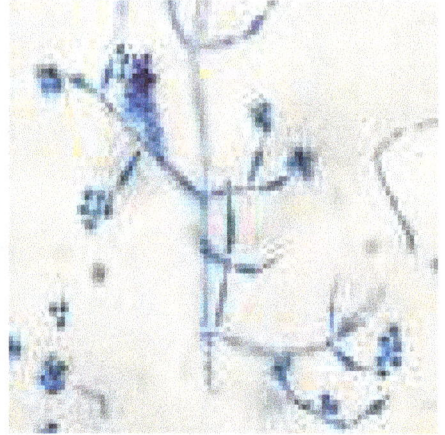

16. *Trichoderma harzianum*

17. *Penicillium citrinum*

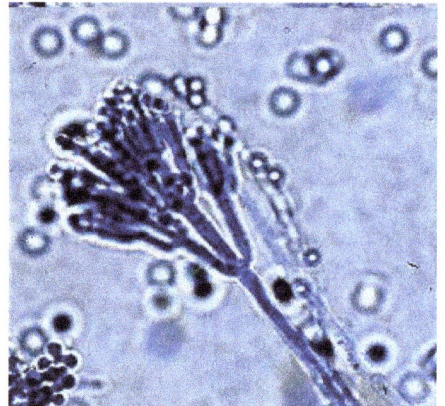

18. *Penicillium oxalicum*

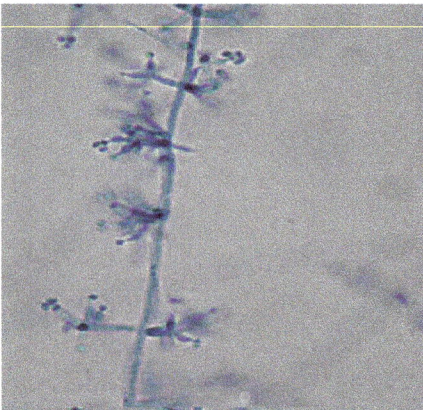

19. *Rhizopus stolonifer*

20. *Trichoderma viride*

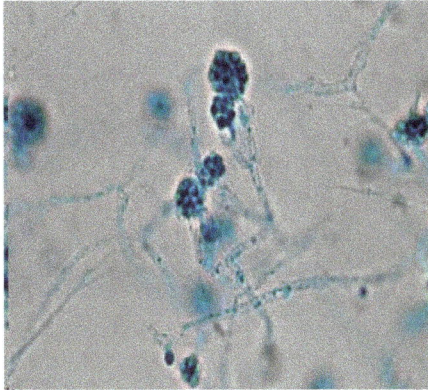

21. *Trichoderma virense*

Figure 20: Fungal species identified from mangrove soils of Bhitarkanika and Mahanadi delta, Odisha (p. 56-59)

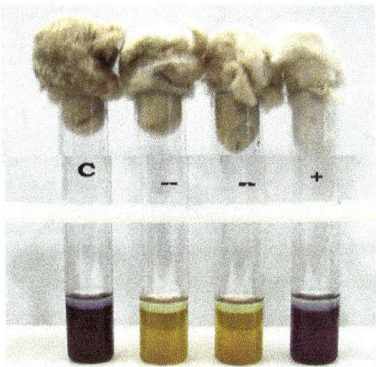

a. Arginine dihydrolase test

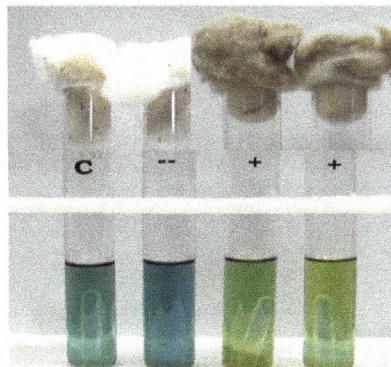

b. Acid-gas production test

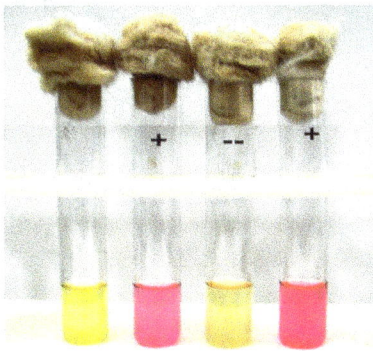

a. Urease test

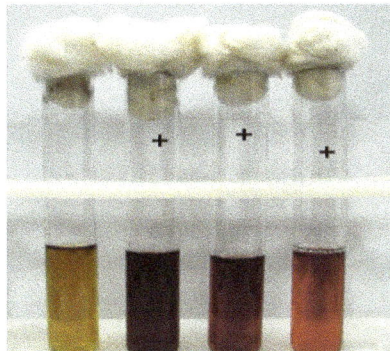

b. Voges Proskauer test

Continue....

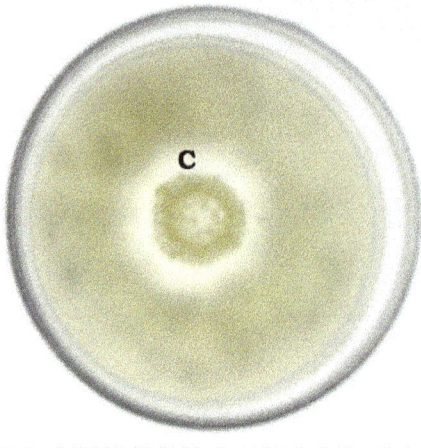

a. Casein hydrolysis   b. Gelatin hydrolysis

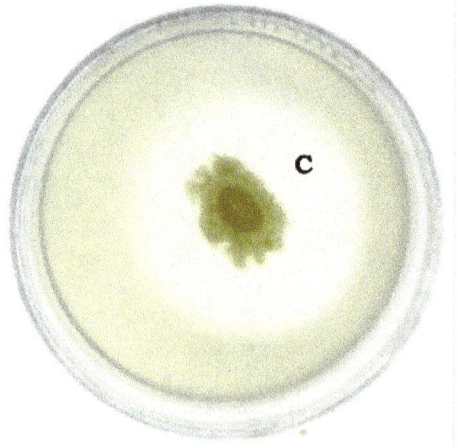

c. Glucose fermentation   d. Starch hydrolysis

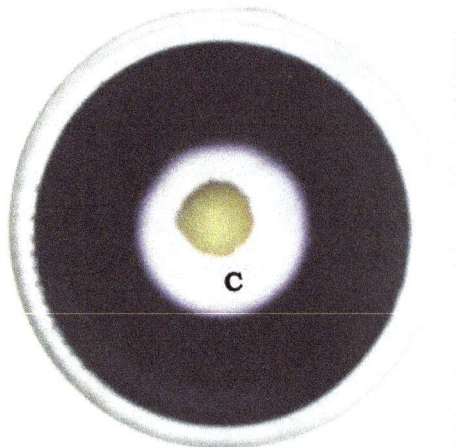

Fig. 23: Some biochemical tests for identification of bacteria (p. 78-79)

Fig. 29: Formation of halozones by 'P' solubilising bacteria on NBRIP-agar medium (p. 89)

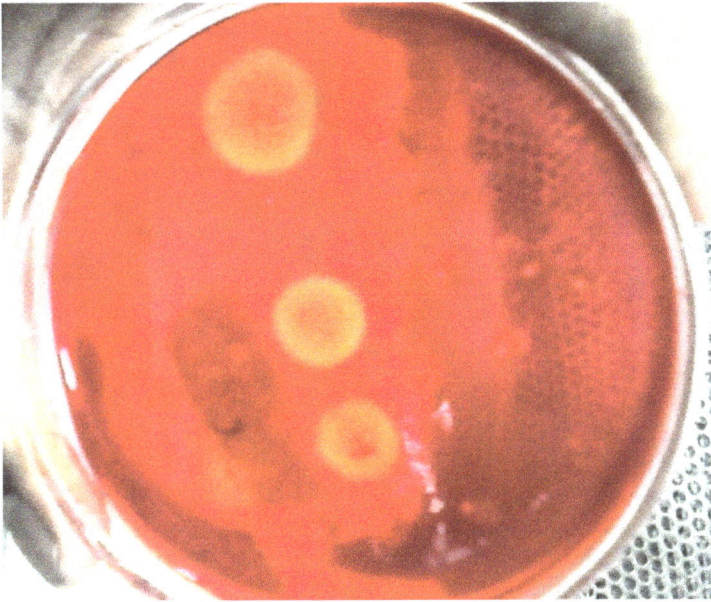

Fig. 34: Clearing zone generated by cellulose degrading bacteria in CM cellulose Congo red agar medium (p. 94)

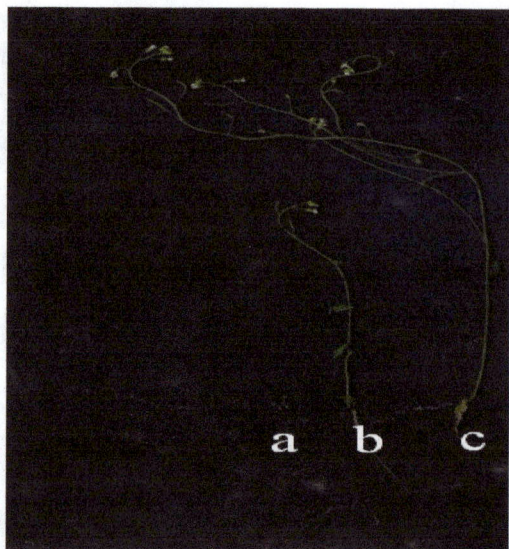

Fig. 48. Effect of *A. faecalis* **on** *A. thaliana* **plant growth promotion. The plantons are (a) modified MSTCP media + plant without bacterial inoculation as negative control, (b) MS media + plant without bacterial inoculation as positive control, (c) modified MSTCP media + plant with 10µl of bacterial inoculation as test planton. (p. 125)**

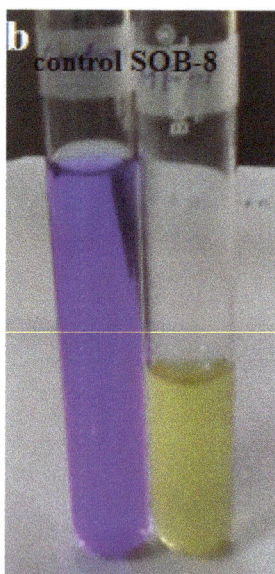

Fig.54: Growth of sulphur oxidising bacteria with control on thiosulphate broth supplied with bromophenol blue as an indicator after 3 days of inoculum (p. 136)

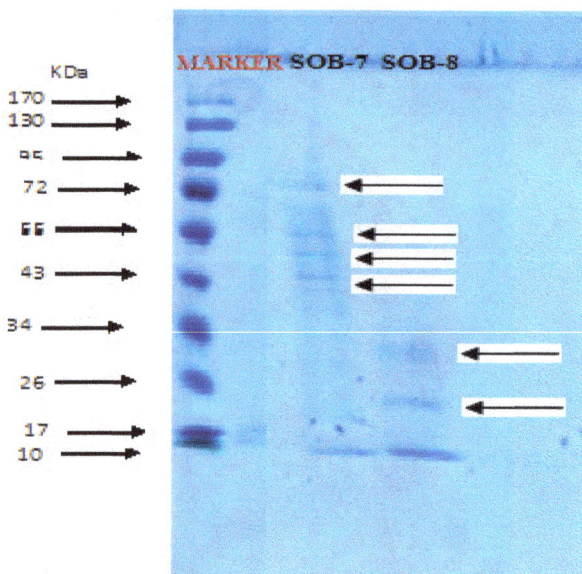

Fig. 60: Partially purified sulphide oxidase profile on SDS PAGE. (p. 145)